Integrated Systems of Meso-Meteorological and Chemical Transport Models

Alexander Baklanov • Alexander Mahura •
Ranjeet S. Sokhi

Integrated Systems of Meso-Meteorological and Chemical Transport Models

Editors
Prof. Alexander Baklanov
Danish Meteorological Institute
Lyngbyvej 100
DK-2100 Copenhagen
Denmark
alb@dmi.dk

Dr. Alexander Mahura
Danish Meteorological Institute
Lyngbyvej 100
DK-2100 Copenhagen
Denmark
ama@dmi.dk

Prof. Ranjeet S. Sokhi
University of Hertfordshire
College Lane
Ctr. Atmospheric & Instrumentation
Research (CAIR)
AL10 9AB Hatfield
United Kingdom
r.s.sokhi@herts.ac.uk

ISBN 978-3-642-13979-6 e-ISBN 978-3-642-13980-2
DOI 10.1007/978-3-642-13980-2
Springer Heidelberg Dordrecht London New York

© Springer-Verlag Berlin Heidelberg 2011
This work is subject to copyright. All rights are reserved, whether the whole or part of the material is concerned, specifically the rights of translation, reprinting, reuse of illustrations, recitation, broadcasting, reproduction on microfilm or in any other way, and storage in data banks. Duplication of this publication or parts thereof is permitted only under the provisions of the German Copyright Law of September 9, 1965, in its current version, and permission for use must always be obtained from Springer. Violations are liable to prosecution under the German Copyright Law.
The use of general descriptive names, registered names, trademarks, etc. in this publication does not imply, even in the absence of a specific statement, that such names are exempt from the relevant protective laws and regulations and therefore free for general use.

Cover design: deblik

Printed on acid-free paper

Springer is part of Springer Science+Business Media (www.springer.com)

Preface

Weather natural hazards, the environment and climate change are of concern to all of us. Especially, it is essential to understand how human activities might impact the nature. Hence, monitoring, research, and forecasting is of the outmost importance. Furthermore, climate change and pollution of the environment do not obey national borders; so, international collaboration on these issues is indeed extremely important.

In the future, the increasing computer power and understanding of physical processes pave the way for developing integrated models of the Earth system and gives a possibility to include interactions between atmosphere, environment, climate, ocean, cryosphere and ecosystems.

Therefore, development of integrated Numerical Weather Prediction (NWP) and Atmospheric Chemical Transport (ACT) models is an important step in this strategic direction and it is a promising way for future atmospheric simulation systems leading to a new generation of models. The EC COST Action 728 "Enhancing Mesoscale Meteorological Modelling Capabilities for Air Pollution and Dispersion Applications" (2004–2009) is aimed at identifying the requirements and propose recommendations for the European strategy for integrated mesoscale NWP-ACT modelling capability.

DMI strongly supports this development. Almost 10 years ago DMI initiated developing an on-line integrated NWP-ACT modelling system, now called Enviro-HIRLAM (Environment – HIgh Resolution Limited Area Model), which includes two-way interactions between meteorology and air pollution for NWP applications and chemical weather forecasting. Recently we also initiated organisation of the Chemical branch in the HIRLAM international consortium (http://hirlam.org), where this model is considered as the baseline model. The Enviro-HIRLAM became an international community model starting January 2009 with several external European organisations joining the research and development team (e.g., from the University of Copenhagen, Denmark; University of Tartu, Estonia; University of Vilnius, Lithuania; Russian State Hydro-Meteorological University; Tomsk State University, Russia; Odessa State Environmental University, Ukraine) with new coming participants.

During 2002–2005, DMI led EC FP5 project FUMAPEX (http://fumapex.dmi.dk), which developed a new generation Integrated Urban Air Quality Information and Forecasting System and implemented such a system in six European cities.

The new EC FP7 project MEGAPOLI (2008–2011) (http://megapoli.info), coordinated by DMI, is also focusing on further developments of integrated systems and studies of interactions between atmospheric pollution from mega cities and meteorological and climatic processes.

These remarks show the importance to organise a workshop to share and analyse international experience in integrated modelling worldwide. The first workshop on "Integration of meteorological and chemical transport models" (http://netfam.fmi.fi/Integ07) was arranged at DMI (Copenhagen, Denmark) on 21–23 May 2007. The workshop was organised in the framework of the COST Action 728 and in cooperation with the Nordic Network on Fine-scale Atmospheric Modelling. Almost 50 participants, including invited experts in integrated modelling and young scientists, from 20 countries attended this event to discuss the experience and further perspectives of coupling air quality and meteorology in fine-scale models. The workshop was aimed at joining both NWP and air quality modellers to discuss and make recommendations on the best practice and strategy for further developments and applications of integrated and coupled modelling systems "NWP and Meso-Meteorology – Atmospheric Chemical Transport". Main emphasis was on fine-resolution models applied for local chemical weather forecasting and considering feedback mechanisms between meteorological and atmospheric pollution (e.g. aerosols) processes. The following topics were in the focus of presentations and discussions:

- Online and offline coupling of meteorological and air quality models
- Implementation of feedback mechanisms, direct and indirect effects of aerosols
- Advanced interfaces between NWP and ACT models
- Model validation studies, including air quality-related episode cases

As a follow-up a young scientist summer school and workshop on "Integrated Modelling of Meteorological and Chemical Transport Processes / Impact of Chemical Weather on Numerical Weather Prediction and Climate Modelling" was organised by DMI and Russian State Hydrometeorological University during 7–15 July 2008 in Russia.

This book, written mostly by invited lectors/speakers of the Copenhagen workshop, is focused on above mentioned workshop topics, summarizes presentations, discussions, conclusions, and provides recommendations. The book is one of the first attempts to give an overall look on such integrated modelling approach. It reviews the current situation with the on-line and off-line coupling of mesoscale meteorological and air quality models around the world (in European countries, USA, Canada, Japan, Australia, etc.) as well as discusses advantages and disadvantages, best practice, and gives recommendations for on-line and off-line coupling of NWP and ACT models, implementation strategy for different feedback mechanisms, direct and indirect effects of aerosols and advanced interfaces between both types of models.

It is my hope that this book will be useful for first of all to those interested in the modelling of meteorology and air pollution, but also for the entire meteorology and atmospheric environment communities, including students, researchers and practical users.

Copenhagen, Denmark DMI Director General, Peter Aakjær

Contents

1 **Introduction – Integrated Systems: On-line and Off-line Coupling of Meteorological and Air Quality Models, Advantages and Disadvantages** 1
Alexander Baklanov

Part I On-Line Modelling and Feedbacks

2 **On-Line Coupled Meteorology and Chemistry Models in the US** 15
Yang Zhang

3 **On-Line Chemistry Within WRF: Description and Evaluation of a State-of-the-Art Multiscale Air Quality and Weather Prediction Model** 41
Georg Grell, Jerome Fast, William I. Gustafson Jr, Steven E. Peckham, Stuart McKeen, Marc Salzmann, and Saulo Freitas

4 **Multiscale Atmospheric Chemistry Modelling with GEMAQ** 55
Jacek Kaminski, Lori Neary, Joanna Struzewska, and John C. McConnell

5 **Status and Evaluation of Enviro-HIRLAM: Differences Between Online and Offline Models** 61
Ulrik Korsholm, Alexander Baklanov, and Jens Havskov Sørensen

6 **COSMO-ART: Aerosols and Reactive Trace Gases Within the COSMO Model** 75
Heike Vogel, D. Bäumer, M. Bangert, K. Lundgren, R. Rinke, and T. Stanelle

7 **The On-Line Coupled Mesoscale Climate–Chemistry Model MCCM: A Modelling Tool for Short Episodes as well as for Climate Periods** 81
Peter Suppan, R. Forkel, and E. Haas

8 **BOLCHEM: An Integrated System for Atmospheric Dynamics and Composition** 89
Alberto Maurizi, Massimo D'Isidoro, and Mihaela Mircea

Part II Off-Line Modelling and Interfaces

9 **Off-Line Model Integration: EU Practices, Interfaces, Possible Strategies for Harmonisation** 97
Sandro Finardi, Alessio D'Allura, and Barbara Fay

10 **Coupling Global Atmospheric Chemistry Transport Models to ECMWF Integrated Forecasts System for Forecast and Data Assimilation Within GEMS** 109
Johannes Flemming, A. Dethof, P. Moinat, C. Ordóñez, V.-H. Peuch, A. Segers, M. Schultz, O. Stein, and M. van Weele

11 **The PRISM Support Initiative, COSMOS and OASIS4** 125
René Redler, Sophie Valcke, and Helmuth Haak

12 **Integrated Modelling Systems in Australia** 139
Peter Manins, M.E. Cope, P.J. Hurley, S.H. Lee, W. Lilley, A.K. Luhar, J.L. McGregor, J.A. Noonan, and W.L. Physick

13 **Coupling of Air Quality and Weather Forecasting: Progress and Plans at met.no** 147
Viel Ødegaard, Leonor Tarrasón, and Jerzy Bartnicki

14 **A Note on Using the Non-hydrostatic Model AROME as a Driver for the MATCH Model** 155
Lennart Robertson and Valentin Foltescu

15 **Aerosol Species in the Air Quality Forecasting System of FMI: Possibilities for Coupling with NWP Models** 159
Mikhail Sofiev and SILAM Team

16 **Overview of DMI ACT-NWP Modelling Systems** 167
Alexander Baklanov, Alexander Mahura, Ulrik Korsholm, Roman Nuterman, Jens Havskov Sørensen, and Bjarne Amstrup

Part III Validation and Case Studies

17 Chemical Modelling with CHASER and WRF/Chem in Japan 181
Masayuki Takigawa, M. Niwano, H. Akimoto, and M. Takahashi

18 Operational Ozone Forecasts for Austria 195
Marcus Hirtl, K. Baumann-Stanzer, and B.C. Krüger

19 Impact of Nesting Methods on Model Performance 201
Ursula Bungert and K. Heinke Schlünzen

20 Running the SILAM Model Comparatively with ECMWF and HIRLAM Meteorological Fields: A Case Study in Lapland ... 207
Marko Kaasik, M. Prank, and M. Sofiev

Part IV Strategy for ACT-NWP Integrated Modeling

21 HIRLAM/HARMONIE-Atmospheric Chemical Transport Models Integration 215
Alexander Baklanov, Sander Tijm, and Laura Rontu

22 Summary and Recommendations on Integrated Modelling 229
Alexander Baklanov, Georg Grell, Barbara Fay, Sandro Finardi, Valentin Foltescu, Jacek Kaminski, Mikhail Sofiev, Ranjeet S. Sokhi, and Yang Zhang

Index 239

List of Contributors

Peter Aakjær Danish Meteorological Institute (DMI), Lyngbyvej 100, DK-2100 Copenhagen, Denmark, paa@dmi.dk

Hajime Akimoto Acid Deposition and Oxidant Research Center, 1182 Sowa Nishi-ku, Nigata-shi 950-2144, Japan, akimoto@adorc.gr.jp

Bjarne Amstrup Danish Meteorological Institute (DMI), Lyngbyvej 100, DK-2100 Copenhagen, Denmark, bja@dmi.dk

Alexander Baklanov Danish Meteorological Institute (DMI), Lyngbyvej 100, DK-2100 Copenhagen, Denmark, alb@dmi.dk

Max Bangert Institut für Meteorologie und Klimaforschung, Karlsruhe Institute of Technology (KIT), Postfach 3640, 76021 Karlsruhe, Germany, max.bangert@kit.edu

Jerzy Bartnicki Norwegian Meteorological Institute (DNMI, met.no), Postboks 43, Blindern 0313, Oslo, Norway, jerzy.bartnicki@met.no

Kathrin Baumann-Stanzer Central Institute for Meteorology and Geodynamics, Hohe Warte 38, 1190 Vienna, Austria, kathrin.baumann-stanzer@zamg.ac.at

Dominique Bäumer Institut für Meteorologie und Klimaforschung, Forschungszentrum, Karlsruhe/Universität Karlsruhe, Postfach 3640, 76021, Karlsruhe, Germany, dominique.baeumer@imk.fzk.de

Ursula Bungert Meteorological Institute, ZMAW, University of Hamburg, Bundesstr. 55, 20146 Hamburg, Germany, ursula.bungert@zmaw.de

Martin E. Cope Commonwealth Scientific and Industrial Research Organization (CSIRO), Marine and Atmospheric Research, PMB 1, Aspendale 3195, VIC, Australia, martin.cope@csiro.au

Alessio D'Allura ARIANET s.r.l, via Gilino 9, 20128, Milano, Italy, a.dallura@aria-net.it

Antje Dethof European Centre for Medium Range Weather Forecast, Shinfield Park, RG2 9AX, Reading, UK, Antje.Inness@ecmwf.int

Massimo D'Isidoro Italian National Agency for New Technologies, Energy and Sustainable Economic Development ENEA, Bologna, Italy; Institute of Atmospheric Sciences and Climate, Italian National Research Council, Rome, Italy, massimo.disidoro@enea.it

Jerome Fast Pacific Northwest National Laboratory, P.O. 999, MSIN K9-30, Richland, WA 99352, USA, jerome.fast@pnl.gov

Barbara Fay German Weather Service (DWD), Frankfurter Str. 135, 63067, Offenbach, Germany, barbara.fay@dwd.de

Sandro Finardi ARIANET s.r.l, via Gilino 9, 20128, Milano, Italy, s.finardi@aria-net.it

Johannes Flemming European Centre for Medium Range Weather Forecast, Shinfield Park, RG2 9AX, Reading, UK, johannes.flemming@ecmwf.int

Valentine Foltescu Swedish Environmental Protection Agency, 106 48 Stockholm, Sweden, valentine.foltescu@naturvardsverket.se

Renate Forkel Institute for Meteorology and Climate Research (IMK-IFU), Karlsruhe Institute of Technology (KIT), Kreuzeckbahnstr. 19, 82467 Garmisch-Partenkirchen, Germany, renate.forkel@kit.edu

Saulo Freitas Center for Weather Forecasting and Climate StudiesINPE, Cachoeira Paulista, Brazil, saulo.freitas@cptec.inpe.br

Georg Grell National Oceanic and Atmospheric Administration (NOAA)/Earth System Research Laboratory (ESRL)/Cooperative Institute for Research in Environmental Sciences (CIRES), 325 Broadway, Boulder CO 80305-3337, USA, Georg.A.Grell@noaa.gov

William I. Gustafson Jr Pacific Northwest National Laboratory, P.O. 999, MSIN K9-30, Richland, WA 99352, USA, william.gustafson@pnl.gov

Helmuth Haak Max Planck Institute for Meteorology, Bundesstrasse 53, 20146, Hamburg, Germany, helmuth.haak@zmaw.de

Edwin Haas Institute for Meteorology and Climate Research (IMK-IFU), Karlsruhe Institute of Technology (KIT), Kreuzeckbahnstr. 19, 82467 Garmisch-Partenkirchen, Germany, edwin.haas@kit.edu

Marcus Hirtl Central Institute for Meteorology and Geodynamics (ZAMG), Hohe Warte 38, 1190 Vienna, Austria, Marcus.Hirtl@zamg.ac.at

Peter J. Hurley Commonwealth Scientific and Industrial Research Organization (CSIRO), Marine and Atmospheric Research, PMB 1, Aspendale 3195, VIC, Australia, peter.hurley@csiro.au

Marko Kaasik Faculty of Science Technology, Institute of Physics, University of Tartu, Tähe 4, 51014 Tartu, Estonia, marko.kaasik@ut.ee

Jacek Kaminski Atmospheric Modelling and Data Assimilation Laboratory, Centre for Research in Earth and Space Science,York University, Toronto, Canada, jwk@wxprime.com

Ulrik Korsholm Danish Meteorological Institute (DMI), Lyngbyvej 100, DK-2100 Copenhagen, Denmark, usn@dmi.dk

Bernd C. Krüger University of Natural Resources and Applied Life Sciences (BOKU), Peter-Jordan-Str. 82, 1190 Vienna, Austria, bernd.krueger@boku.ac.at

Sun Hee Lee Commonwealth Scientific and Industrial Research Organization (CSIRO), Marine and Atmospheric Research, PMB 1, Aspendale 3195, VIC, Australia, sunhee.lee@csiro.au

W. Lilley Commonwealth Scientific and Industrial Research Organization (CSIRO), Marine and Atmospheric Research, PMB 1, Aspendale 3195, VIC, Australia, bill.lilley@csiro.au

Ashok K. Luhar Commonwealth Scientific and Industrial Research Organization (CSIRO), Marine and Atmospheric Research, PMB 1, Aspendale 3195, VIC, Australia, ashok.luhar@csiro.au

Kristina Lundgren Institut für Meteorologie und Klimaforschung, Karlsruhe Institute of Technology (KIT), Postfach 3640, 76021 Karlsruhe, Germany, kristina.lundgren@kit.edu

Alexander Mahura Danish Meteorological Institute (DMI), Lyngbyvej 100, DK-2100 Copenhagen, Denmark, ama@dmi.dk

Peter Manins Commonwealth Scientific and Industrial Research Organization (CSIRO), Marine and Atmospheric Research, PMB 1, Aspendale 3195, VIC, Australia, peter.manins@csiro.au

Alberto Maurizi Institute of Atmospheric Sciences and Climate, Italian National Research Council, Bologna, Italy, a.maurizi@isac.cnr.it

John C. McConnell Atmospheric Modelling and Data Assimilation Laboratory, Centre for Research in Earth and Space ScienceYork University, 4700 Keele Street, Toronto ON, M3J 1P3, Canada, jcmcc@yorku.ca

John L. McGregor Commonwealth Scientific and Industrial Research Organization (CSIRO), Marine and Atmospheric Research, PMB 1, Aspendale 3195, VIC, Australia, john.mcgregor@csiro.au

Stuart McKeen National Oceanic and Atmospheric Administration (NOAA)/ Earth System Research Laboratory (ESRL)/Cooperative Institute for Research in Environmental Sciences (CIRES), 325 Broadway, Boulder CO 80305-3337, USA, stuart.a.mckeen@noaa.gov

Mihaela Mircea Italian National Agency for New Technologies, Energy and Sustainable Economic Development ENEA, Bologna, Italy; Institute of Atmospheric Sciences and Climate, Italian National Research Council, Rome, Italy, mihaela.mircea@enea.it

Philippe Moinat CNRM-GAME, Météo-France and CNRS URA, 357, 42 avenue G. Coriolis, 31057, Toulouse, France, Philippe.Moinat@cnrm.meteo.fr

Lori Neary Atmospheric Modelling and Data Assimilation Laboratory Centre for Research in Earth and Space Science, York University, 4700 Keele Street, Toronto ON, M3J 1P3, Canada, lori@yorku.ca

Masaaki Niwano Sumitomo Chemical, 4-2-1 Takatsukasa, Takaraduka Hyogo 665-8555, Japan, niwanom@sc.sumitomo-chem.co.jp

Julie A. Noonan Commonwealth Scientific and Industrial Research Organization (CSIRO), Marine and Atmospheric Research, PMB 1, Aspendale 3195, VIC, Australia, julie.noonan@csiro.au

Roman Nuterman Danish Meteorological Institute (DMI), Lyngbyvej 100, DK-2100 Copenhagen, Denmark, ron@dmi.dk

Viel Ødegaard Norwegian Meteorological Institute (DNMI, met.no), Postboks 43, Blindern, 0313 Oslo, Norway, v.odegaard@met.no

Carlos Ordóñez Laboratoire d'Aérologie, 14 avenue Edouard Belin, 31400, Toulouse, France, carlos.ordonez@metoffice.gov.uk

Steven E. Peckham National Oceanic and Atmospheric Administration (NOAA)/ Earth System Research Laboratory (ESRL)/Cooperative Institute for Research in Environmental Sciences (CIRES), 325 Broadway, Boulder, CO 80305-3337, USA, steven.peckham@noaa.gov

Vincent-Henri Peuch CNRM-GAME, Météo-France and CNRS URA, 1357, 42 avenue G. Coriolis, 31057, Toulouse, France, Vincent-Henri.Peuch@meteo.fr

W.L. Physick Commonwealth Scientific and Industrial Research Organization (CSIRO), Marine and Atmospheric Research, PMB 1, Aspendale 3195, VIC, Australia, bill.physick@csiro.au

Marje Prank Finnish Meteorological Institute, Erik Palmenin aukio 1, P.O. Box 503, 00101 Helsinki, Finland, marje.prank@fmi.fi

René Redler NEC Laboratories Europe – IT Division, NEC Europe Ltd., Sankt Augustin, Germany, rene.redler@zmaw.de

Rayk Rinke Institut für Meteorologie und Klimaforschung, Karlsruhe Institute of Technology (KIT), Postfach 3640, 76021 Karlsruhe, Germany, rayk.rinke@kit.edu

Lennart Robertson Swedish Meteorological and Hydrological Institute (SMHI), SE-601 76 Norrköping, Sweden, lennart.robertson@smhi.se

Laura Rontu Finish Meteorological Institute (FMI), P.O. Box 503, 00101 Helsinki, Finland, laura.rontu@fmi.fi

Marc Salzmann Atmospheric and Oceanic Sciences Program, Princeton University, Princeton, NJ, USA, Marc.Salzmann@noaa.gov

K. Heinke Schlünzen Meteorological Institute, KlimaCampus, University of Hamburg, Bundesstr. 55, 20146 Hamburg, Germany, heinke.schluenzen@zmaw.de

Martin Schultz ICG-2 Research Center Juelich, Wilhelm-Johnen-Str, 52425 Juelich, Germany, m.schultz@fz-juelich.de

Arjo Segers TNO, Princetonlaan 6, 3584 CB, Utrecht, The Netherlands, Arjo.Segers@tno.nl

SILAM Team Finnish Meteorological Institute (FMI), P.O. Box 503, 00101 Helsinki, Finland

Mikhail Sofiev Finnish Meteorological Institute (FMI), P.O. Box 503, 00101 Helsinki, Finland, mikhail.sofiev@fmi.fi

Ranjeet S. Sokhi Centre for Atmospheric and Instrumentation Research (CAIR) University of Hertfordshire College Lane, Hatfield AL10 9AB, UK, r.s.sokhi@herts.ac.uk

Jens H. Sørensen Danish Meteorological Institute (DMI), Lyngbyvej 100, DK-2100 Copenhagen, Denmark, jhs@dmi.dk

Tanja Stanelle Institut für Meteorologie und Klimaforschung, Forschungszentrum, Karlsruhe/Universität Karlsruhe, Postfach 3640, 76021, Karlsruhe, Germany, tanja.stanelle@imk.fzk.de

Olaf Stein ICG-2 Research Center Juelich, Wilhelm-Johnen-Str, 52425, Juelich, Germany, o.stein@fz-juelich.de

Joanna Struzewska Faculty of Environmental EngineeringWarsaw University of Technology, Nowowiejska 20, 00-653, Warsaw, Poland, joanna.struzewska@is.pw.edu.pl

Peter Suppan Institute for Meteorology and Climate Research (IMK-IFU), Karlsruhe Institute of Technology (KIT), Garmisch-Partenkirchen, Germany, peter.suppan@kit.edu

Masaaki Takahashi Center for Climate System Research, University of Tokyo, 5-1-5 Kashiwanoha, Kashiwa 277-8568, Japan, masaaki@ccsr.u-tokyo.ac.jp

Masayuki Takigawa Japan Agency for Marine-Earth Science and Technology, 3173-25 Showa-machi, Kanazawa-ku, Yokohama, Kanagawa 236-0001, Japan, takigawa@jamstec.go.jp

Leonor Tarrasón Norwegian Institute for Air research (NILU), Postboks 100, 2027 Kjeller, Norway, lta@nilu.no

Sander Tijm Royal Netherlands Meteorological Institute (KNMI), Postbus 201, 3730 AE De Bilt, The Netherlands, tijm@knmi.nl

Sophie Valcke CERFACS, 42 Av. Coriolis, 31057 Toulouse, France, Sophie.Valcke@cerfacs.fr

Heike Vogel Institut für Meteorologie und Klimaforschung, Karlsruhe Institute of Technology (KIT), Postfach 3640, 76021 Karlsruhe, Germany, heike.vogel@kit.edu

Michiel van Weele Royal Netherlands Meteorological Institute (KNMI), P.O. Box 201, 3730 AE De Bilt, The Netherlands, weelevm@knmi.nl

Yang Zhang Department of Marine, Earth and Atmospheric Sciences, North Carolina State University, Raleigh NC 27695, USA, yang_zhang@ncsu.edu

Chapter 1
Introduction – Integrated Systems: On-line and Off-line Coupling of Meteorological and Air Quality Models, Advantages and Disadvantages

Alexander Baklanov

1.1 Introduction

Historically air pollution forecasting and numerical weather prediction (NWP) were developed separately. This was plausible in the previous decades when the resolution of NWP models was too poor for meso-scale air pollution forecasting. Due to modern NWP models approaching meso- and city-scale resolution (due to advances in computing power) and the use of land-use databases and remote sensing data with finer resolution, this situation is changing. As a result the conventional concepts of meso- and urban-scale air pollution forecasting need revision along the lines of integration of meso-scale meteorological models (MetMs) and atmospheric chemical transport models (ACTMs). For example, a new Environment Canada conception suggests to switch from weather forecasting to environment forecasting. Some European projects (e.g. FUMAPEX, see: fumapex.dmi.dk) already work in this direction and have set off on a promising path. In case of FUMAPEX it is the Urban Air Quality Information and Forecasting Systems (UAQIFS) integrating NWP models, urban air pollution (UAP) and population exposure models (Baklanov et al. 2007b), see Fig. 1.1.

In perspective, integrated NWP-ACTM modelling may be a promising way for future atmospheric simulation systems leading to a new generation of models for improved meteorological, environmental and "chemical weather" forecasting.

Both, off-line and on-line coupling of MetMs and ACTMs are useful in different applications. Thus, a timely and innovative field of activity will be to assess their interfaces, and to establish a basis for their harmonization and benchmarking. It will consider methods for the aggregation of episodic results, model down-scaling as well as nesting. The activity will also address the requirements of meso-scale meteorological models suitable as input to air pollution models.

The COST728 Action (http://www.cost728.org) addressed key issues concerning the development of meso-scale modelling capabilities for air pollution

A. Baklanov
Danish Meteorological Institute (DMI), Lyngbyvej 100, DK-2100 Copenhagen, Denmark
e-mail: alb@dmi.dk

A. Baklanov et al. (eds.), *Integrated Systems of Meso-Meteorological and Chemical Transport Models*, DOI 10.1007/978-3-642-13980-2_1,
© Springer-Verlag Berlin Heidelberg 2011

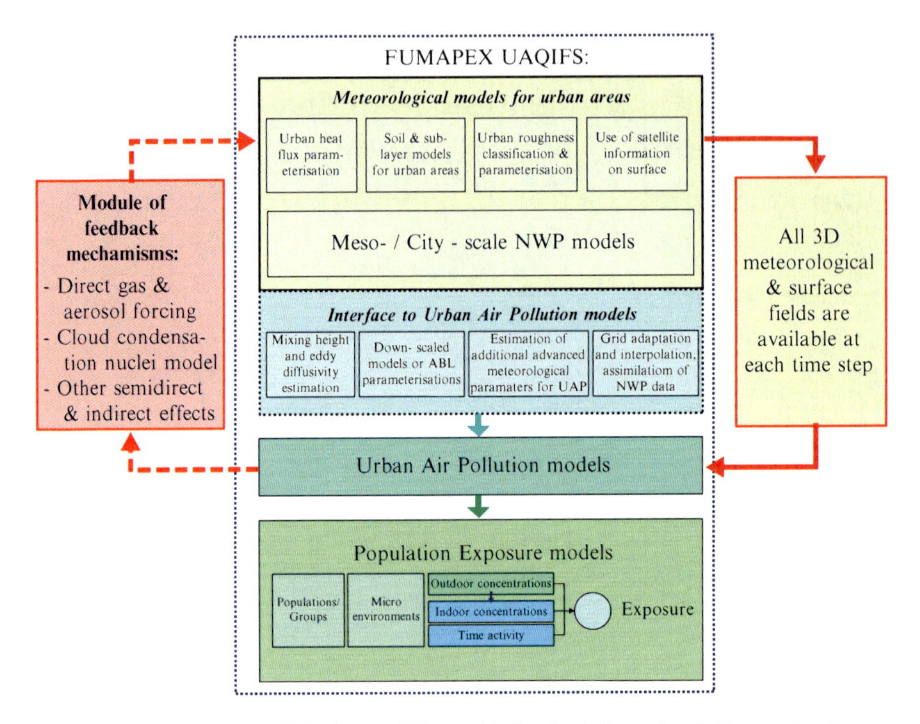

Fig. 1.1 Extended FUMAPEX scheme of Urban Air Quality Information & Forecasting System (UAQIFS) including feedbacks. Improvements of meteorological forecasts (NWP) in urban areas, interfaces and integration with UAP and population exposure models following the off-line or on-line integration (Baklanov 2005; after EMS-FUMAPEX 2005)

and dispersion applications and, in particular, it encouraged the advancement of science in terms of integration methodologies and strategies in Europe. The final integration strategy will not be focused around any particular model; instead it will be possible to consider an open integrated system with fixed architecture (module interface structure) and with a possibility of incorporating different MetMs/NWP models and ACTMs. Such a strategy may only be realised through jointly agreed specifications of module structure for easy-to-use interfacing and integration.

The overall aim of working group 2 (WG2) of the COST 728 Action, "Integrated systems of MetM and ACTM: strategy, interfaces and module unification", is to identify the requirements for the unification of MetM and ACTM modules and to propose recommendations for a European strategy for integrated meso-scale modelling capabilities. The first report of WG2 (Baklanov et al. 2007a) compiles existing state-of-the-art methodologies, approaches, models and practices for building integrated (off-line and on-line) meso-scale systems in different, mainly European, countries. The report also includes an overview and a summary of existing integrated models and their characteristics as they are presently used. The model contributions were compiled using COST member contributions, each focusing on national model systems.

1.2 Methodology for Model Integration

The modern strategy for integrating MetMs and ACTMs is suggested to incorporate air quality modelling as a combination of (at least) the following factors: air pollution, regional/urban climate/meteorological conditions and population exposure. This combination is reasonable due to the following facts: meteorology is the main source of uncertainty in air pollution and emergency preparedness models, meteorological and pollution components have complex and combined effects on human health (e.g., hot spots in Paris, July 2003), pollutants, especially aerosols, influence climate forcing and meteorological events (such as, precipitation and thunderstorms).

In this context, several levels of MetM and ACTM coupling/integration can be considered:

Off-Line

- Separate ACTMs driven by meteorological input data from meteo-preprocessors, measurements or diagnostic models
- Separate ACTMs driven by analysed or forecasted meteodata from NWP archives or datasets
- Separate ACTMs reading output-files from operational NWP models or specific MetMs at limited time intervals (e.g. 1, 3, 6 h)

On-Line

- On-line access models, when meteodata are available at each time-step (possibly via a model interface as well)
- On-line integration of ACTM into MetM, where feedbacks may be considered. We will use this definition for on-line coupled/integrated modelling

The main advantages of the On-line coupled modelling approach comprise:

- Only one grid is employed and no interpolation in space is required
- There is no time interpolation
- Physical parametrizations and numerical schemes (e.g. for advection) are the same; No inconsistencies
- All 3D meteorological variables are available at the right time (each time step)
- There is no restriction in variability of meteorological fields
- Possibility exists to consider feedback mechanisms, e.g. aerosol forcing
- There is no need for meteo- pre/post-processors.

However, the on-line approach is not always the best way of the model integration. For some specific tasks (e.g., for emergency preparedness, when NWP data are available) the off-line coupling is more efficient way.

The main advantages of Off-line models comprise:

- There is the possibility of independent parametrizations
- They are more suitable for ensembles activities
- They are easier to use for the inverse modelling and adjoint problem
- There is the independence of atmospheric pollution model runs on meteorological model computations
- There is more flexible grid construction and generation for ACTMs
- This approach is suitable for emission scenarios analysis and air quality management.

The on-line integration of meso-scale meteorological models and atmospheric aerosol and chemical transport models enables the utilisation of all meteorological 3D fields in ACTMs at each time step and the consideration of two-way feedbacks between air pollution (e.g. urban aerosols), meteorological processes and climate forcing. These integration methodologies have been demonstrated by several of the COST action partners such as the Danish Meteorological Institute, with the DMI-ENVIRO-HIRLAM model (Chenevez et al. 2004; Baklanov et al. 2004, 2008; Korsholm et al. 2007) and the COSMO consortium with the Lokal Modell (Vogel et al. 2006; Wolke et al. 2003).

These model developments will lead to a new generation of integrated models for: climate change modelling, weather forecasting (e.g., in urban areas, severe weather events, etc.), air quality, long-term assessments of chemical composition and chemical weather forecasting (an activity of increasing importance which is supported by a new COST action ES0602 started in 2007).

1.3 Overview of European On-Line Integrated Models

The experience from other European, as well as non-European union communities, will need to be integrated. On our knowledge on-line coupling was first employed at the Novosibirsk scientific school (Marchuk 1982; Penenko and Aloyan 1985; Baklanov 1988), for modelling active artificial/anthropogenic impacts on atmospheric processes. Currently American, Canadian and Japanese institutions develop and use on-line coupled models operationally for air quality forecasting and for research (GATOR-MMTD: Jacobson, 2005, 2006; WRF-Chem: Grell et al. 2005; GEM-AQ: Kaminski et al. 2005).

Such activities in Europe are widely dispersed and a COST Action seems to be the best approach to integrate, streamline and harmonize these national efforts towards a leap forward for new breakthroughs beneficial for a wide community of scientists and users.

Such a model integration should be realized following a joint elaborated specification of module structure for potential easy interfacing and integration. It might develop into a system, e.g. similar to the USA ESMF (Earth System Modelling Framework, see e.g.: Dickenson et al. 2002) or European PRISM (PRogram for Integrating Earth System Modelling) specification for integrated Earth System Models: http://prism.enes.org/ (Valcke et al. 2006).

Community Earth System Models (COSMOS) is a major international project (http://cosmos.enes.org) involving different institutes in Europe, in the US and in Japan, for the development of complex Earth System Models (ESM). Such models are needed to understand large climate variations of the past and to predict future climate changes.

The main differences between the COST-728 integrating strategy for meso-scale models and the COSMOS integration strategy regards the spatial and temporal scales. COSMOS is focusing on climate time-scale processes, general (global and regional) atmospheric circulation models and atmosphere, ocean, cryosphere and biosphere integration, while the meso-scale integration strategy will focus on forecast time-scales of 1 to 4 days and omit the cryosphere and the larger temporal and spatial scales in atmosphere, ocean and biosphere.

The COST728 model overview (Baklanov et al. 2007a) shows a surprisingly large (at least ten) number of on-line coupled MetM and ACTM systems already being used in Europe (see also more information in Table 1.1):

- BOLCHEM (CNR ISAC, Italy)
- ENVIRO-HIRLAM (DMI, Denmark)
- LM-ART (Inst. for Meteorology and Climate Research (IMK-TRO), KIT, Germany)
- LM-MUSCAT (IfT Leipzig, Germany)
- MCCM (Inst. for Meteorology and Climate Research (IMK-IFU), KIT, Germany)
- MESSy: ECHAM5 (MPI-C Mainz, Germany)
- MC2-AQ (York Univ, Toronto, University of British Columbia, Canada, and Warsaw University of Technology, Poland)
- GEM/LAM-AQ (York Univ, Toronto, University of British Columbia, Canada, and Warsaw University of Technology, Poland)
- WRF-CHem: Weather Research and Forecast and Chemistry Community modelling system (NCAR and many other organisations)
- MESSy: ECHAM5-Lokalmodell LM planned at MPI-C Mainz, Univ. of Bonn, Germany

However, it is necessary to mention, that many of the above on-line models were not build for the meso-meteorological scale, and several of them (GME, MESSy) are global-scale modelling systems, originating from the climate modelling community. Besides, at the current stage most of the on-line coupled models do not consider feedback mechanisms or include only simple direct effects of aerosols on meteorological processes (COSMO LM-ART and MCCM). Only two meso-scale on-line integrated modelling systems (WRF-Chem and ENVIRO-HIRLAM) consider feedbacks with indirect effects of aerosols.

Table 1.1 On-line coupled MetM – ACTMs (Baklanov et al. 2007a)

Model name	On-line coupled chemistry	Time step for coupling	Feedback
BOLCHEM	Ozone as prognostic chemically active tracer		None
ENVIRO-HIRLAM	Gas phase, aerosol and heterogeneous chemistry	Each HIRLAM time step	Yes
WRF-Chem	RADM+Carbon Bond, Madronich+Fast-J photolysis, modal+sectional aerosol	Each model time step	Yes
COSMO LM-ART	Gas phase chem (58 variables), aerosol physics (102 variables), pollen grains	Each LM time step	Yes[a]
COSMO LM-MUSCAT[b]	Several gas phase mechanisms, aerosol physics	Each time step or time step multiple	None
MCCM	RADM and RACM, photolysis (Madronich), modal aerosol	Each model time step	(Yes)[a]
MESSy: ECHAM5	Gases and aerosols		Yes
MESSy: ECHAM5-COSMO LM (planned)	Gases and aerosols		Yes
MC2-AQ	Gas phase: 47 species, 98 chemical reactions and 16 photolysis reactions	Each model time step	None
GEM/LAM-AQ	Gas phase, aerosol and heterogeneous chemistry	Set up by user – in most cases every time step	None
Operational ECMWF model (IFS)	Prog. stratos passive O3 tracer	Each model time step	Yes
ECMWF GEMS modelling	GEMS chemistry		
GME	Progn. stratos passive O3 tracer	Each model time step	
OPANA=MEMO +CBMIV		Each model time step	

[a]Direct effects only
[b]On-line access model

1.4 Feedback Mechanisms, Aerosol Forcing in Meso-meteorological Models

In a general sense air quality and ACTM modelling is a natural part of the climate change and MetM/NWP modelling process. The role of greenhouse gases (such as water vapour, CO_2, O_3 and CH_4) and aerosols in climate change has been highlighted as a key area of future research (Watson et al. 1997; IPCC 2007, 2001; AIRES 2001). Uncertainties in emission projections of gaseous pollutants and aerosols (especially secondary organic components) need to be addressed urgently to advance our understanding of climate forcing (Semazzi 2003). In relation to aerosols, their diverse sources, complex physicochemical characteristics and large spatial gradients make their role in climate forcing particularly challenging to quantify. In addition to primary emissions, secondary particles, such as, nitrates, sulphates and organic compounds, also result from chemical reactions involving precursor gases such as SO_x, DMS, NO_x, volatile organic compounds and oxidising agents including ozone. One consequence of the diverse nature of aerosols is that they exhibit negative (e.g. sulphates) as well as positive (e.g. black carbon) radiative forcing characteristics (IPCC 2007, 2001; Jacobson 2002). Although much effort has been directed towards gaseous species, considerable uncertainties remain in size dependent aerosol compositional data, physical properties as well as processes controlling their transport and transformation, all of which affect the composition of the atmosphere (Penner et al. 1998; Shine 2000; IPCC 2007, 2001). Probably one of the most important sources of uncertainty relates to the indirect effect of aerosols as they also contribute to multiphase and microphysical cloud processes, which are of considerable importance to the global radiative balance (Semazzi 2003).

In addition to better parameterisations of key processes, improvements are required in regional and global scale atmospheric modelling (IPCC 2005; Semazzi 2003). Resolution of regional climate information from atmosphere-ocean general circulation models remains a limiting factor. Vertical profiles of temperature, for example, in climate and air quality models need to be better described. Such limitations hinder the prospect of reliably distinguishing between natural variability (e.g. due to natural forcing agents, solar irradiance and volcanic effects) and human induced changes caused by emissions of greenhouse gases and aerosols over multidecadal timescales (Semazzi 2003). Consequently, the current predictions of the impact of air pollutants on climate, air quality and ecosystems or of extreme events are unreliable (e.g. Watson et al. 1997). Therefore it is very important in the future research to address all the key areas of uncertainties so as provide an improved modelling capability over regional and global scales and an improved integrated assessment methodology for formulating mitigation and adaptation strategies.

In this concern one of the important tasks is to develop a modelling instrument of coupled "Atmospheric chemistry/Aerosol" and "Atmospheric Dynamics/Climate" models for integrated studies, which is able to consider the feedback mechanisms, e.g. aerosol forcing (direct and indirect) on the meteorological processes and climate change (see Fig. 1.2).

Chemical species influencing weather and atmospheric processes include greenhouse gases which warm near-surface air and aerosols such as sea salt, dust, primary and secondary particles of anthropogenic and natural origin. Some aerosol particle components (black carbon, iron, aluminium, polycyclic and nitrated aromatic compounds) warm the air by absorbing solar and thermal-IR radiation, while others (water, sulphate, nitrate, most of organic compounds) cool the air by backscattering incident short-wave radiation to space.

It is necessary to highlight those effects of aerosols and other chemical species on meteorological parameters have many different pathways (such as, direct, indirect and semi-direct effects) and they have to be prioritized and considered in

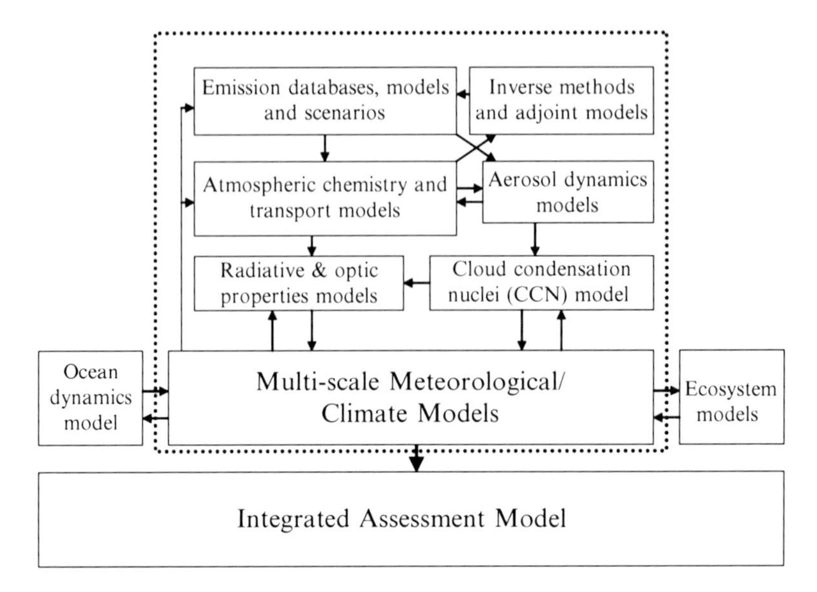

Fig. 1.2 The integrated system structure for studies of the meso-scale meteorology and air pollution, and their interaction

on-line coupled modelling systems. Following Jacobson (2002) the following effects of aerosol particles on meteorology and climate can be distinguished:

- Self-Feedback Effect
- Photochemistry Effect
- Smudge-Pot Effect
- Daytime Stability Effect
- Particle Effect through Surface Albedo
- Particle Effect through Large-Scale Meteorology
- Indirect Effect
- Semi-direct Effect
- BC-Low-Cloud-Positive Feedback Loop

Sensitivity studies are needed to understand the relative importance of different feedback mechanisms. Implementation of the feedbacks into integrated models could be realized in different ways with varying complexity. The following variants serve as examples:

One-Way Integration (Off-Line)

- The chemical composition fields from ACTMs may be used as a driver for Regional/Global Climate Models, including aerosol forcing on meteorological processes. This strategy could also be realized for NWP or MetMs.

Two-Way Integration

- Driver and partly aerosol feedbacks, for ACTMs or for NWP (data exchange with a limited time period); off-line or on-line access coupling, with or without the following iterations with corrected fields)
- Full chain of two-way interactions, feedbacks included on each time step (on-line coupling/integration)

For the realization of all aerosol forcing mechanisms in integrated systems it is necessary to improve not only ACTMs, but also NWP/MetMs. The boundary layer structure and processes, including radiation transfer, cloud development and precipitation must be improved. Convection and condensation schemes need to be adjusted to take the aerosol–cloud microphysical interactions into account, and the radiation scheme needs to be modified to include the aerosol effects.

1.5 Concluding Remarks

The on-line integration of meso-scale meteorological models and atmospheric aerosol and chemical transport models enables the utilization of all meteorological 3D fields in ACTMs at each time step and the consideration of the feedbacks of air pollution (e.g. urban aerosols) on meteorological processes and climate forcing.

Developments in on-line coupled modelling will lead to a new generation of integrated models for climate change modelling, weather forecasting (e.g., in urban areas, severe weather events, etc.), air quality, long-term assessment chemical composition and chemical weather forecasting.

Main advantages of the on-line modelling approach include:

- Only one grid; No interpolation in space;
- No time interpolation;
- Physical parametrizations are the same; No inconsistencies;
- All 3D meteorological variables are available at the right time (each time step);
- No restriction in variability of meteorological fields;
- Possibility to consider feedback mechanisms;
- Does not need meteo- pre/post-processors.

While the main advantages of the off-line approach include:

- Possibility of independent parametrizations
- More suitable for ensemble activities
- Easier to use for the inverse modelling and adjoint problem
- Independence of atmospheric pollution model runs on meteorological model computations
- More flexible grid construction and generation for ACTMs
- Suitable for emission scenarios analysis and air quality management

The COST728 model overview shows a quite surprising number of on-line coupled MetM and ACTM systems already being used in Europe. However, many of the on-line coupled models were not built for the meso-meteorological scale, and they (e.g. GME, ECMWF C-IFS, MESSy) are global-scale modelling systems and first of all designed for climate change modelling. Besides, at the current stage most of the on-line coupled models do not consider feedback mechanisms or include only direct effects of aerosols on meteorological processes (like COSMO LM-ART and MCCM). Only two meso-scale on-line integrated modelling systems (mentioned in the COST 728 list), namely WRF-Chem and ENVIRO-HIRLAM, consider feedbacks with indirect effects of aerosols.

Acknowledgements This study is supported by the COST Action 728, EU FP7 MEGAPOLI and Danish CEEH projects. The author is grateful to a number of COST728, FUMAPEX and DMI colleagues, who participated in the above-mentioned projects, for productive collaboration and discussions.

References

AIRES (2001) AIRES in ERA, European Commission, EUR 19436

Baklanov A (1988) Numerical modelling in mine aerology. USSR Academy of Science, Apatity, 200 p, (in Russian)

Baklanov A (2005) Meteorological advances and systems for urban air quality forecasting and assessments. Short Papers of the 5th international conference on urban air quality. Valencia, Spain, 29–31 Mar 2005, CLEAR, pp 22–25

Baklanov A, Gross A, Sørensen JH (2004) Modelling and forecasting of regional and urban air quality and microclimate. J Comput Technol 9:82–97

Baklanov A, Fay B, Kaminski J, Sokhi R (2007a) Overview of existing integrated (off-line and on-line) meso-scale systems in Europe. COST728 WG2 Deliverable 2.1 Report, May 2007. WMO-COST publication. GAW Report No. 177, 107 p. Available from: http://www.cost728.org

Baklanov A, Hänninen O, Slørdal LH, Kukkonen J, Bjergene N, Fay B, Finardi S, Hoe SC, Jantunen M, Karppinen A, Rasmussen A, Skouloudis A, Sokhi RS, Sørensen JH, Ødegaard V (2007b) Integrated systems for forecasting urban meteorology, air pollution and population exposure. Atmos Chem Phys 7:855–874

Baklanov A, Korsholm U, Mahura A, Petersen C, Gross A (2008) EnviroHIRLAM: on-line coupled modelling of urban meteorology and air pollution. Adv Sci Res 2:41–46

Chenevez J, Baklanov A, Sørensen JH (2004) Pollutant transport schemes integrated in a numerical weather prediction model: Model description and verification results. Meteorol Appl:11 (3):265–275

COSMOS: Community earth system models integrating strategy. Web-site: http://cosmos.enes.org

Dickenson RE, Zebiak SE, Anderson JL, Blackmon ML, DeLuca C, Hogan TF, Iredell M, Ji M, Rood R, Suarez MJ, Taylor KE (2002) How can we advance our weather and climate models as a community? Bull Am Met Soc 83:431–434

EMS-FUMAPEX (2005) Urban Meteorology and Atmospheric Pollution. In: Baklanov A, Joffre S, Galmarini S (eds) Atmos Chem Phys J (Special Issue)

Grell GA, Peckham SE, Schmitz R, McKeen SA, Frost G, Skamarock WC, Eder B (2005) Fully coupled "online" chemistry within the WRF model. Atmos Environ 39(37):6957–6975

IPCC (2005) IPCC expert meeting on emission estimation of aerosols relevant to climate change. Geneva, Switzerland, 2–4 May 2005

IPCC (2007) Climate change 2007: the physical science basis. Contribution of Working Group I to the Fourth Assessment Report on the Intergovernmental Panel on Climate Change. Cambridge University Press, Cambridge

Korsholm U, Baklanov A, Gross A, Sørensen JH (2007) Influence of offline coupling interval on meso-scale representations. Atmos Environ 43:4805–4810

Jacobson MZ (2002) Atmospheric pollution: history, science and regulation. Cambridge University Press, New York

Jacobson MZ (2005) Fundamentals of atmospheric modelling, 2nd edn. Cambridge University Press, New York, 813 pp

Jacobson MZ (2006) Comment on "Fully coupled 'online' chemistry within the WRF model," by Grell et al., Atmos Environ 39:6957–697

Marchuk GI (1982) Mathematical modelling in the environmental problems. Nauka, Moscow

Penenko VV, Aloyan AE (1985) Models and methods for environment protection problems. Nauka, Novosibirsk, (in Russian)

Penner JE et al (1998) Climate forcing by carbonaceous and sulphate aerosols. Clim Dyn 14:839–851

Semazzi F (2003) Air quality research: perspective from climate change modelling research. Environ Int 29:253–261

Shine KP (2000) Radiative forcing of climate change. Space Sci Rev 94:363–373

Valcke S, Guilyardi E, Larsson C (2006) PRISM and ENES: a European approach to Earth system modelling. Concurr Comput: Pract Exper 18:231–245

Vogel B, Hoose C, Vogel H, Kottmeier Ch (2006) A model of dust transport applied to the Dead Sea area. Meteorol Z 14:611–624

Watson RT et al (1997) The regional impacts of climate change: an assessment of vulnerability. Special Report for the Intergovernmental Panel on Climate Change Working Group II. Cambridge University Press. Cambridge, UK

Wolke R, Hellmuth O, Knoth O, Schröder W, Heinrich B, Renner E (2003) The chemistry-transport modelling system LM-MUSCAT: description and CITYDELTA applications. Proceedings of the 26th international technical meeting on air pollution and its application. Istanbul, May 2003, pp 369–379

Part I
On-Line Modelling and Feedbacks

Chapter 2
On-Line Coupled Meteorology and Chemistry Models in the US

Yang Zhang

2.1 Introduction

The climate–chemistry–aerosol–cloud–radiation feedbacks are important in the context of many areas including climate modelling, air quality (AQ)/atmospheric chemistry modelling, numerical weather prediction (NWP) and AQ forecasting, as well as integrated atmospheric-ocean-land surface modelling at all scales. Some potential impacts of aerosol feedbacks include a reduction of downward solar radiation (direct effect); a decrease in surface temperature and wind speed but an increase in relative humidity and atmospheric stability (semi-direct effect), a decrease in cloud drop size but an increase in drop number via serving as cloud condensation nuclei (first indirect effect), as well as an increase in liquid water content, cloud cover, and lifetime of low level clouds but a suppression of precipitation (the second indirect effect). Aerosol feedbacks are traditionally neglected in meteorology and AQ modelling due largely to historical separation of meteorology, climate, and AQ communities as well as our limited understanding of underlying mechanisms. Those feedbacks, however, are important as models accounting (e.g., Jacobson 2002; Chung and Seinfeld 2005) or not accounting (e.g., Penner 2003) for those feedbacks may give different results and future climate changes may be affected by improved air quality. Accurately simulating those feedbacks requires fully-coupled models for meteorological, chemical, physical processes and presents significant challenges in terms of both scientific understanding and computational demand. In this work, the history and current status of development and application of on-line models are reviewed. Several representative models developed in the US are used to illustrate the current status of on-line coupled models. Major challenges and recommendations for future development and improvement of on-line- coupled models are provided.

Y. Zhang
Department of Marine, Earth and Atmospheric Sciences, North Carolina State University, Raleigh, NC 27695, USA
e-mail: yang_zhang@ncsu.edu

A. Baklanov et al. (eds.), *Integrated Systems of Meso-Meteorological and Chemical Transport Models*, DOI 10.1007/978-3-642-13980-2_2,
© Springer-Verlag Berlin Heidelberg 2011

2.2 History of Coupled Chemistry/Air Quality and Climate/Meteorology Models

2.2.1 Concepts and History of On-Line Models

Atmospheric chemistry/air quality and climate/meteorology modelling was traditionally separated prior to mid. 1970s. The three-dimensional (3D) atmospheric chemical transport models (ACTMs) until that time were primarily driven by either measured/analyzed meteorological fields or outputs at a time resolution of 1–6 h from a mesoscale meteorological model on urban/regional scale or outputs at a much coarser time resolution (e.g., 6-h or longer) from a general circulation model (GCM) (referred to as off-line coupling). In addition to a large amount of data exchange, this off-line separation does not permit simulation of feedbacks between AQ and climate/meteorology and may result in an incompatible and inconsistent coupling between both meteorological and AQ models and a loss of important process information (e.g., cloud formation and precipitation) that occur at a time scale smaller than that of the outputs from the off-line climate/meteorology models. Such feedbacks, on the other hand, are allowed in the fully-coupled on-line models, without space and time interpolation of meteorological fields but commonly with higher computational costs.

The earliest attempt in coupling global climate/meteorology and chemistry can be traced back to late 1960s, when 3D transport of ozone and simple stratospheric chemistry (e.g., the Chapman reactions, the NO_x catalytic cycle, and reactions between hydrogen and atomic oxygen) was first incorporated into a GCM to simulate global ozone (O_3) production and transport (e.g., Hunt 1969; Clark 1970; Cunnold et al. 1975; Schlesimger and Mintz 1979). In such models, atmospheric transport and simple stratospheric O_3 chemistry are simulated in one model, accounting for the effect of predicted O_3 on radiation heating and the effect of radiation heating on atmospheric circulation, which in turn affects distribution of O_3. Since mid. 1980s, a large number of on-line global climate/chemistry models have been developed to address the Antarctic/stratospheric O_3 depletion (e.g., Cariolle et al. 1990; Cariolle and Deque 1986; Rose and Brasseur 1989; Austin et al. 1992; Rasch et al. 1995; Jacobson 1995), tropospheric O_3 and sulfur cycle (e.g., Roelofs and Lelieveld 1995; Feichter et al. 1996; Barth et al. 2000), tropospheric aerosol and its interactions with cloud (e.g., Chuang et al. 1997; Lohmann et al. 2000; Jacobson 2000, 2001a; Easter et al. 2004). The coupling in most on-line models, however, has been enabled only for very limited prognostic gaseous species such as O_3 and/or bulk aerosol (e.g., Schlesimger and Mintz 1979) or selected processes such as transport and gas-phase chemistry (i.e., incompletely- or partially-coupling). This is mainly because such a coupling largely restricts to gas-phase/heterogeneous chemistry and simple aerosol/cloud chemistry and microphysics and often neglects the feedbacks between prognostic chemical species (e.g., O_3 and aerosols) and radiation (e.g., Roelofs and Lelieveld 1995;

Eckman et al. 1996; Barth et al. 2000) and aerosol indirect effects (e.g., Liao et al. 2003), with a few exceptions after mid. 1990s when truly-coupled systems were developed to enable a full range of feedbacks between meteorology/climate variables and a myriad of gases and size-resolved aerosols (e.g., Jacobson 1995, 2000; Ghan et al. 2001a, b, c).

The earliest attempt in coupling meteorology and air pollution in mesoscale models can be traced back to early 1980s (Baklanov et al. 2007 and references therein). Since then, a number of mesoscale on-line coupled meteorology-chemistry models have been developed in North America (e.g., Jacobson 1994, 1997a, b; Mathur et al. 1998; Côté et al. 1998; Grell et al. 2000) and Australia (e.g., Manins 2007) but mostly developed recently by European researchers largely through the COST Action 728 (http://www.cost728.org) (e.g., Baklanov et al. 2004, 2007, and references therein). The coupling was enabled between meteorology and tropospheric gas-phase chemistry only in some regional models (e.g., Grell et al. 2000); and among more processes/components including meteorology, chemistry, aerosols, clouds, and radiation (e.g., Jacobson 1994, 1997a, b; Jacobson et al. 1996; Mathur et al. 1998; Grell et al. 2005; Fast et al. 2006; Zhang et al. 2005a, b, 2010a, b; Krosholm et al. 2007; and Misenis and Zhang 2010). Similar to global models, a full range of climate–chemistry–aerosol–cloud–radiation feedbacks is treated in very few mesoscale models (e.g., Jacobson 1994, 1997a, b; Grell et al. 2005).

Two coupling frameworks are conventionally used in all mesoscale and global on-line coupled models: one couples a meteorology model with an AQ model in which the two systems operate separately but exchange information every time step through an interface (referred to as separate on-line coupling), the other integrates an AQ model into a meteorology model as a unified model system in which meteorology and AQ variables are simulated together in one time step without a model-to-model interface (referred to as unified on-line coupling). Transport of meteorological and chemical variables is typically simulated with separate schemes in separate on-line models but the same scheme in unified on-line models. Depending on the objectives of the applications, the degrees of coupling and complexities in coupled atmospheric processes in those models vary, ranging from a simple coupling of meteorology and gas-phase chemistry (e.g., Rasch et al. 1995; Grell et al. 2000) to the most sophistic coupling of meteorology, chemistry, aerosol, radiation, and cloud (e.g., Jacobson 1994, 2004b, 2006; Grell et al. 2005). While on-line coupled models can in theory enable a full range of feedbacks among major components and processes, the coupling is typically enabled in two modes: partially-coupled where only selected species (e.g., O_3) and/or processes (e.g., transport and gas-phase chemistry) are coupled and other processes (e.g., solar absorption of O_3 and total radiation budget) remain decoupled; fully-coupled where all major processes are coupled and a full range of atmospheric feedbacks can be realistically simulated. At present, very few fully-coupled on-line models exist; and most on-line models are partially-coupled and still under development.

2.2.2 History of Representative On-Line Models in the US

In this review, five models on both regional and global scales developed in the US are selected to represent the current status of on-line-coupled models. These include:

- One global-through-urban model, i.e., the Stanford University's Gas, Aerosol, TranspOrt, Radiation, General Circulation, Mesoscale, Ocean Model (GATOR/GCMOM) (Jacobson 2001c, 2002, 2004a; Jacobson et al. 2004)
- One mesoscale model, i.e., the National Oceanic and Atmospheric Administration (NOAA)'s Weather Research Forecast model with Chemistry (WRF/Chem) (Grell et al. 2005; Fast et al. 2006; Zhang et al. 2010a)
- Three global models, i.e., the National Center for Atmospheric Chemistry (NCAR)'s Community Atmospheric Model v. 3 (CAM3), the Pacific Northwest National laboratory (PNNL)'s Model for Integrated Research on Atmospheric Global Exchanges *version 2* (MIRAGE2) (Textor et al. 2006; Ghan and Easter 2006), and the Caltech unified GCM (Liao et al. 2003; Liao and Seinfeld 2005)

All these models predict gases, aerosols, and clouds with varying degrees of complexities in chemical mechanisms and aerosol/cloud microphysics. The history and current status of these models along with other relevant models are reviewed below.

Jacobson (1994, 1997a, b) and Jacobson et al. (1996) developed the first unified fully-coupled on-line model that accounts for major feedbacks among meteorology, chemistry, aerosol, cloud, radiation on urban/regional scales: a gas, aerosol, transport, and radiation AQ model/a mesoscale meteorological and tracer dispersion model (GATOR/MMTD, also called GATORM). Grell et al. (2000) developed a unified on-line coupled meteorology and gas-phase chemistry model: Multiscale Climate Chemistry Model (MCCM, also called MM5/Chem). Built upon MM5/Chem and NCAR's WRF, Grell et al. (2002) developed a unified fully-coupled on-line model, WRF/Chem, to simulate major atmospheric feedbacks among meteorology, chemistry, aerosol, and radiation. This is the first community on-line model in the US. Since its first public release in 2002, WRF/Chem has attracted a number of external developers and users from universities, research organizations, and private sectors to continuously and collaboratively develop, improve, apply, and evaluate the model. In WRF/Chem, transport of meteorological and chemical variables is treated using the same vertical and horizontal coordinates and the same physics parametrization with no interpolation in space and time. In addition to Regional Acid Deposition Model v.2 (RADM2) in MM5/Chem, WRF/Chem includes an additional gas-phase mechanism: the Regional Atmospheric Chemistry Mechanism (RACM) of Stockwell et al. (1997) and a new aerosol module: the Modal Aerosol Dynamics Model for Europe (MADE) (Ackermann et al. 1998) with the secondary organic aerosol model (SORGAM) of Schell et al. (2001) (referred to as MADE/SORGAM). Two additional gas-phase mechanisms and two new aerosol modules have been recently incorporated into WRF/Chem by external developers (Fast et al. 2006; Zhang et al. 2005a, b, 2007, 2010a; Pan et al. 2008). The two new gas-phase mechanisms are the Carbon-Bond Mechanism version Z (CBMZ) (Zaveri and Peters 1999) and the 2005 version of Carbon Bond mechanism (CB05) of Yarwood et al. (2005). The two new aerosol modules are the Model for Simulating Aerosol Interactions and Chemistry (MOSAIC) (Zaveri et al. 2008) and the Model of Aerosol Dynamics, Reaction, Ionization, and Dissolution (MADRID) (Zhang et al. 2004, 2010c).

On a global scale, a number of climate or AQ models have been developed in the past three decades among which very few of them are on-line models. Since its initial development as a general circulation model without chemistry, CCM0 (Washington 1982), the NCAR's Community Climate Model (CCM) has evolved to be one of the first unified on-line climate/chemistry models, initially with gas-phase chemistry only (e.g., CCM2 (Rasch et al. 1995) and CCM3 (Kiehl et al. 1998; Rasch et al. 2000)) and most recently with additional aerosol treatments (e.g., CAM3 (Collins et al. 2004, 2006a, b; and CAM4 (http://www.ccsm.ucar.edu)). Jacobson (1995, 2000, 2001a) developed a unified fully-coupled Gas, Aerosol, TranspOrt, Radiation, and General circulation model (GATORG) built upon GATORM and a 1994 version of the University of Los Angeles GCM (UCLA/GCM). Jacobson (2001b, c) linked the regional GATORM and global GATORG and developed the first unified, nested global-through-urban scale Gas, Aerosol, Transport, Radiation, General Circulation, and Mesoscale Meteorological model, GATOR/GCMM. GATOR/GCMM was designed to treat gases, size- and composition-resolved aerosols, radiation, and meteorology for applications from the global to urban (<5 km) scales and accounts for radiative feedbacks from gases, size-resolved aerosols, liquid water and ice particles to meteorology on all scales. GATOR/GCMM was extended to Gas, Aerosol, TranspOrt, Radiation, General Circulation, Mesoscale, Ocean Model (GATOR/GCMOM) in Jacobson (2004a, 2006) and Jacobson et al. (2004, 2006). Built upon NCAR CCM2 and PNNL Global Chemistry Model (GChM), MIRAGE1 was developed and can be run off-line or fully-coupled on-line (Ghan et al. 2001a, b, c and Easter et al. 2004). In MIRAGE2, the gas/aerosol treatments are an integrated model imbedded in NCAR CAM2 (i.e. unified on-line coupling). Several on-line-coupled global climate/aerosol models with full oxidant chemistry have also been developed since early 2000 but most of them do not include all feedbacks, in particular, aerosol indirect effects; and they are under development (e.g., Liao et al. 2003). Among all 3D models that have been developed for climate and AQ studies at all scales, GATOR/GCMOM, MIRAGE, and WRF/Chem represent the state of science global and regional coupled models; and GATOR/GCMOM appears to be the only model that represents gas, size- and composition-resolved aerosol, cloud, and meteorological processes from the global down to urban scales via nesting, allowing feedback from gases, aerosols, and clouds to meteorology and radiation on all scales in one model simulation.

2.3 Current Treatments in On-Line Coupled Models in the US

In this section, model features and treatments for the five representative on-line coupled meteorology and chemistry models developed in the US are reviewed in terms of model systems and typical applications, aerosol and cloud properties, aerosol and cloud microphysics and aerosol–cloud interactions. As shown in Table 2.1, four out of the five models are unified on-line models (i.e., GATOR/GCMOM; WRF/Chem, CAM3, and Caltech unified GCM) and one (i.e., MIRAGE) is a separate on-line model, all with different levels of details in gas-phase chemistry and aerosol and cloud treatments ranging from the simplest one in CAM3 to the most complex one in GATOR/GCMOM. Those models have been developed for different applications. As shown in Table 2.2, the treatments of aerosol properties in those models are different in terms of composition, size distribution, aerosol mass/number concentrations, mixing state, hygroscopicity, and radiative properties. For example, MIRAGE2 treats the least number of species, and GATOR/GCMOM treats the most. Size distribution of all aerosol components are prescribed in Caltech unified GCM and that of all aerosols except sea-salt and dust is prescribed in CAM3; they are predicted in the other three models. Prescribed aerosol size distribution may introduce significant biases in simulated aerosol direct and indirect radiative forcing that highly depends on aerosol size distributions. The mixing state of aerosols affects significantly the predictions of direct/indirect radiative forcing. The internally-mixed (i.e., well-mixed) hydrophilic treatment for BC is unphysical and reality lies between the externally-mixed, hydrophobic and core treatments. Among the five models, GATOR/GCMOM is the only model treating internal/external aerosol mixtures with a coated BC core. All the five models predict aerosol mass concentration, but only some of them can predict aerosol number concentration (e.g., GATOR/GCMOM, WRF/Chem, and MIRAGE2). For aerosol radiative properties, GATOR/GCMOM assumes a BC core surrounded by a shell where the refractive indices (RIs) of the dissolved aerosol components are determined from partial molar refraction theory and those of the remaining aerosol components are calculated to be volume-averaged based on core-shell MIE theory. MIRAGE2, WRF/Chem, and Caltech unified GCM predict RIs and optical properties using Mie parametrizations that are function of wet surface mode radius and wet RI of each mode. Volume mixing is assumed for all components, including insoluble components. The main difference between Caltech unified GCM and MIRAGE2 (and WRF/Chem) is that Caltech unified GCM prescribes size distribution, but MIRAGE2 predicts it. In CAM3, RIs and optical properties are prescribed for each aerosol type, size, and wavelength of the external mixtures.

Table 2.3 summarizes model treatments of cloud properties, reflecting the levels of details in cloud microphysics treatments from the simplest in Caltech unified GCM to the most sophistic in GATOR/GCMOM. GATOR/GCMOM uses prognostic, multiple size distributions (typically three, for liquid, ice, and graupel), each with 30 size sections. MIRAGE2 and WRF/Chem simulate bulk condensate in single size distribution, with either a modal distribution (MIRAGE2) or a sectional

Table 2.1 Model systems and typical applications of on-line models

Model system/scale	Met. model	Chemical transport model (main features)	Typical applications	Example References
GATOR/GCMOM & Predecessors (Global-through-urban)	MMTD GCMM GCMOM	Gas-phase chemistry: CBM-EX: (247 reactions, 115 species); bulk or size-resolved aqueous-phase sulfate, nitrate, organics, chlorine, oxidant, radical chemistry (64 kinetic reactions); size-resolved, prognostic aerosol/cloud with complex processes	Current/future met/chem/rad feedbacks; Direct/indirect effects; AQ/health effect	Jacobson (1994, 1997a, b, 2001c, 2002, 2004a, b), Jacobson et al. (2004, 2007), Jacobson (2006)
WRF/Chem (Mesoscale)	WRF	RADM2, RACM, CBMZ, CB05 (156–237 reactions, 52–77 species); bulk aqueous-phase RADM chemistry (MADE/SORGAM) or CMU mechanism (MOSAIC/MADRID); Three aerosol modules (MADE/SORGAM, MOSAIC, and MADRID with size/mode-resolved, prognostic aerosol/cloud treatments	Forecast/hindcast, Met/chem. feedbacks; O_3, $PM_{2.5}$; aerosol direct and indirect effects	Grell et al. (2005), Fast et al. (2006), Zhang et al. (2005a, b, 2007, 2010a, b), Tie et al. (2007), Misenis and Zhang (2010)
CAM3 & Predecessors (Global)	CCM3/CCM2/CCM1	Sulfur chemistry (14 reactions) prescribed CH_4, N_2O, CFCs/MOZART4 gas-phase chemistry (167 reactions, 63 species); Sulphur chemistry; Bulk aqueous-phase sulfate chemistry of S(IV) (4 equilibria and 2 kinetic reactions); prognostic aerosol/cloud treatments with prescribed size distribution	Climate; direct/indirect effects; hydrological cycle	Rasch et al. (1995), Kiehl et al. (1998), Collins et al. (2004, 2006a, b)

(continued)

Table 2.1 (continued)

Model system/scale	Met. model	Chemical transport model (main features)	Typical applications	Example References
MIRAGE2 & 1 (Global)	CAM2/CCM2	Gas-phase CO–CH4-oxidant chem. (MIRAGE1 only); bulk aqueous-phase sulfate chemistry (6 equilibria and 3 kinetic reactions); Mode-resolved simple aerosol treatment; Prognostic aerosol/cloud treatments	Trace gases and PM; Sulphur cycle; Direct/indirect effects	Ghan et al. (2001a, b, c), Zhang et al. (2002), Easter et al. (2004), Textor et al. (2006), Ghan and Easter, (2006)
Caltech unified GCM (Global)	GISS GCM II'	Harvard tropospheric O_3-NO_x-hydrocarbon chemistry (305–346 reactions, 110–225 species); bulk aqueous-phase chemistry of S(IV) (5 equilibria and 3 kinetic reactions); prognostic aerosol/cloud treatments with prescribed size distribution	Global chemistry–aerosol interactions; aerosol direct radiative forcing; the role of heterogeneous chemistry; impact of future climate change on O_3 and aerosols	Liao et al. (2003), Liao and Seinfeld (2005)

Table 2.2 Treatments of aerosol properties of on-line models

Model system	Composition	Size distribution	Aerosol mixing state	Aerosol mass/number	Aerosol hygroscopicity	Aerosol radiative properties
GATOR/GCMOM	47 species (sulfate, nitrate, ammonium, BC, OC, sea-salt, dust, water, crustal)	Sectional (17–30): variable, multiple size distributions	A coated core, internal/external mixtures	Predicted/predicted	Simulated hydrophobic-to-hydrophilic conversion for all aerosol components	Simulated volume-average refractive indices and optical properties based on core-shell MIE theory
WRF/Chem	Sulfate, nitrate, ammonium, BC, OC, water in all 3 aerosol modules, sea-salt, and carbonate in MOSAIC/MADRID	Modal (3): variable (MADE/SORGAM); Sectional (8): variable (MOSAIC/MADRID); single size distribution	Internal	Predicted/Predicted	similar to MIRAGE2	similar to MIRAGE2
CAM3	Sulfate, nitrate, ammonium, BC, OC, sea-salt, dust, water	Modal (4): predicted dust and sea-salt, prescribed other aerosols; single size distribution	External	Predicted/Predicted or predicted/Diagnosed from mass	hydrophobic and hydrophilic BC/OC with a fixed conversion rate	Prescribed RI and optical properties for each aero. type, size, and wavelength, for external mixtures
MIRAGE2	Sulfate, BC, OC, sea-salt, dust, water	Modal (4): variable; single size distribution	Externally mixed modes with internal mixtures within each mode	Predicted/Diagnosed or predicted	Simulated BC/OC with prescribed hygroscopicities for OC and dust	Parameterized RI and optical properties based on wet radius and RI of each mode

(*continued*)

Table 2.2 (continued)

Model system	Composition	Size distribution	Aerosol mixing state	Aerosol mass/number	Aerosol hygroscopicity	Aerosol radiative properties
Caltech unified GCM (Global)	Sulfate, nitrate, ammonium, BC, OC, sea-salt, dust, water, Ca^{2+}	Sectional (11) prescribed for sea-salt; Sectional (6) prescribed for mineral dust; Modal (1): prescribed size distribution for other aerosols; single size distribution for all aerosols	BC, OC, and mineral dust externally mixed with internally-mixed SO_4^{2-}, NH_4^+, NO_3^-, sea-salt, and H_2O; different aerosol mixing states for chemistry and radiative forcing calculation	Predicted aerosol mass; aerosol number not included	Simulated BC/OC with prescribed hygroscopicities	Simulated optical properties based on Mie theory with size- and wavelength-dependent refractive indices

Table 2.3 Treatments of cloud properties of on-line models

Model system	Hydrometeor types in clouds	Cloud droplet size distribution	Cloud droplet number	CCN/IDN composition	CCN/IDN spectrum	Cloud radiative properties
GATOR/GCMOM	Size-resolved liquid, ice, graupel, aerosol core components, in stratiform subgrid convective clouds	Prognostic, sectional (30), multiple size distributions (3)	Prognostic, size- and composition-dependent from multiple aerosol size distributions	All types of aerosols treated for both CCN/IDN	Predicted with Köhler theory; sectional (13–17); multiple size distributions (1–16) for both CCN/IDN	Simulated volume-average refractive indices and optical properties based on MIE theory and a dynamic effective medium approximation
WRF/Chem	Bulk water vapour, rain, snow, cloud ice, cloud water, graupel or a subset of them depending on microphysics schemes used in both stratiform and subgrid convective clouds	Prognostic, sectional, single size distribution (MOSAIC)	similar to MIRAGE2 (MOSAIC)	similar to MIRAGE2 but sectional; CCN only	similar to MIRAGE2 but sectional, CCN only	similar to MIRAGE2 but sectional (MOSAIC)
CAM3	Bulk liquid and ice in both stratiform and subgrid convective clouds	Prognostic in microphysics calculation but prescribed in sedimentation and radiation calculation as a function of temperature by phase and location	similar to MIRAGE2	All treated species except hydrophobic species; CCN only	Prescribed; CCN only	similar to MIRAGE2

(*continued*)

Table 2.3 (continued)

Model system	Hydrometeor types in clouds	Cloud droplet size distribution	Cloud droplet number	CCN/IDN composition	CCN/IDN spectrum	Cloud radiative properties
MIRAGE2	Bulk liquid and ice in both stratiform and subgrid convective clouds	Prognostic, modal, single size distribution	Prognostic, aerosol size- and composition-dependent, parameterized	All treated species; CCN only	Function of aerosol size and hygroscopicity based on Köhler theory; CCN only	Prognostic, parameterized in terms of cloud water, ice mass, and number
Caltech unified GCM (Global)	Bulk liquid and ice in both stratiform and subgrid convective clouds	Diagnosed from predicted cloud water content; single size distribution	constant cloud droplet number based on observations	None	None	Simulated based on MIE theory with different parametrizations for liquid and ice clouds

distribution (WRF/Chem/MOSAIC). CAM3 treats bulk liquid and ice with the same prognostic droplet size treatment as MIRAGE2. Caltech unified GCM treats bulk liquid and ice with their distributions diagnosed from predicted cloud water content. Among the five models, Caltech unified GCM is the only model that prescribes cloud droplet number, which is predicted in the other four models. CAM3, MIRAGE2, and WRF/Chem use the same treatment for droplet number, with droplet nucleation parameterized by Abdul-Razzak and Ghan (2000). GATOR treats prognostic, size- and composition-dependent cloud droplet number from multiple aerosol size distributions. While an empirical relationship between sulfate aerosols and CCN is commonly used in most atmospheric models, CCN is calculated from Köhler theory using the aerosol size distribution and hygroscopicity in all models but Caltech unified GCM. Other than Caltech unified GCM that does not treat CCN and Ice Deposition Nuclei (IDN), all other four models treat the competition among different aerosol species for CCN but the hydrophobic species are not activated in CAM3 since it assumes an external-mixture. Among the five models, GATOR/GCMOM is the only model that simulates composition of IDN. MIARGE and CAM use a prognostic parametrization in terms of cloud water, ice mass, and number to predict cloud radiative properties. WRF/Chem also uses the same method but with sectional approach. Caltech unified GCM simulates cloud optical properties based on MIE theory and prescribed Gamma distribution for liquid clouds. GATOR/GCMOM simulates volume-average cloud RIs and optical properties based on MIE theory and an iterative dynamic effective medium approximation (DEMA) to account for multiple BC inclusions within clouds. The DEMA is superior to classic effective-medium approximation that is used by several mixing rules such as the volume-average RI mixing rule (Jacobson 2006).

Table 2.4 shows model treatments of aerosol chemistry and microphysics that differ in many aspects. For example, Caltech unified GCM treats aerosol thermodynamics only, the rest of models treat both aerosol thermodynamics and dynamics such as coagulation and new particle formation via homogeneous nucleation. The degree of complexity varies in terms of number of species and reactions treated and assumptions made in the inorganic aerosol thermodynamic modules used in those models. The simplest module, MARS-A, is used in WRF/Chem/MADE/SORGAM, and the most comprehensive module, EQUISOLV II, is used in GATOR/GCMOM. For secondary organic aerosol (SOA) formation, both CAM3 and MIRAGE2 use prescribed aerosol yields for a few condensable volatile organic compounds (VOCs), which is the simplest, computationally most efficient approach but it does not provide a mechanistic understanding of SOA formation. GATOR/GCMOM simulates SOA formation from 10 to 40 classes VOCs via condensation and dissolution based on Henry's law. Caltech unified GCM simulates SOA formation based on a reversible absorption of five classes of biogenic VOCs and neglects that from anthropogenic VOCs. In MADE/SORGAM in WRF/Chem, SOA formation via reversible absorption of eight classes VOCs is simulated based on Caltech smog-chamber data. Two approaches are used to simulate SOA formation in WRF/Chem/MADRID (Zhang et al. 2004). MADRID 1 uses an absorptive approach for 14 parent VOCs and 38 SOA species. MADRID 2 combines

Table 2.4 Treatments of aerosol chemistry and microphysics of on-line models

Model system	Inorganic aero. thermodynamic equilibrium	Secondary organic aerosol formation	New particle formation	Condensation of gases on aerosols	Coagulation	Gas/particle mass transfer
GATOR/GCMOM	EQUISOLV II, major inorganic salts and crustal species	Condensation; dissolution based on Henry's law (10–40 classes VOCs)	Binary homogeneous nucleation of H_2SO_4 and H_2O, Ternary nucleation, T- and RH-dependent	Dynamic condensation of all condensible species based on growth law (e.g., H_2SO_4, VOCs) using the Analytical Predictor of Condensation (APC) with the moving center scheme	Sectional, multiple size distributions, accounts for van der Waals and viscous forces, and fractal geometry	Dynamic approach with a long time step (150–300 s) (PNG/EQUISOLV II) for all treated species
WRF/Chem	MARS-A (SORGAM); MESA-MTEM; (MOSAIC); ISORROPIA; (MADRID)	Reversible absorption (8 classes VOCs) based on smog-chamber data (SORGAM), Absorption (MADRID1) and combined absorption and dissolution (MADRID2), no SOA treatment in MOSAIC	Binary homogeneous nucleation of H_2SO_4 and H_2O; T- and RH-dependent; sectional; different eqs. in different aero modules	Dynamic condensation of H_2SO_4 and VOCs using the modal approach (SORGAM), of H_2SO_4, MSA, and NH_3 using the Adaptive Step Time-split Explicit Euler Method (ASTEEM) method (MOSAIC), and of volatile inorganic species using the	Modal/Sectional (MADE/SORGAM, MOSAIC), single size distribution, fine-mode only	1. Full equilibrium for HNO_3 and NH_3 in MADE/SORGAM and all species in MADRID 2. Dynamic for H_2SO_4 in MADE/SORGAM; Dynamic for all species in MOSAIC and MADRID 3. Hybrid in MADRID

				APC with the moving center scheme (MADRID)		
CAM3	MOZART4 with regime equili. for sulfate, nitrate, and ammonium	Prescribed SOA yield for α-pinene, n-butane, and toluene	None	Instantaneous condensation of inorganic species	None	Full equilibrium involving $(NH_4)_2SO_4$ and NH_4NO_3
MIRAGE2	$(NH_4)_2SO_4$, no nitrate	Prescribed SOA yield	Binary homogeneous nucleation of H_2SO_4 and H_2O; T- and RH-dependent	Dynamic condensation of H_2SO_4 and MSA based on Fuchs and Sutugin growth law	Modal, single size distribution, fine-mode only; Brownian diffusion	Dynamical approach for H_2SO_4 and MSA
Caltech unified GCM (Global)	ISORROPIA with regime equili. for sulfate, nitrate, ammonium, sea-salt, and water	Reversible absorption for 5 biogenic SVOC classes	None	None	None	Full equilibrium involving $(NH4)_2SO_4$ and NH_4NO_3

absorption and dissolution approaches to simulate an external mixture of 42 hydrophilic and hydrophobic VOCs. SOA formation in not treated in MOSAIC. Coagulation is currently not treated in CAM3 but simulated with a modal approach in MIRAGE2, sectional approach in GATOR/GCMOM, and both in WRF/Chem/MADE/SORGAM and MOSIAC. Different from other model treatments, GATOR accounts for van der Waals, viscous forces, and fractal geometry in simulating coagulation among particles from multiple size distributions (Jacobson and Seinfeld 2004). For gas/particle mass transfer, CAM3 and Caltech unified GCM use the simplest full equilibrium approach. MIRAGE2 uses a dynamic approach for H_2SO_4 and MSA. GATOR/GCMOM uses a computationally-efficient dynamic approach with a long time step (150–300 s) (PNG/EQUISOLV II) for all treated species (Jacobson 2005). In WRF/Chem, a full equilibrium approach is used for HNO_3 and NH_3 in MADE/SORGAM, a dynamic approach is used in MOSAIC. MADRID offers three approaches: full equilibrium, dynamic, and hybrid; their performance has been evaluated in Zhang et al. (1999, 2010a) and Hu et al. (2008). Hu et al. (2008) have shown that the bulk equilibrium approach is computationally-efficient but less accurate, whereas the kinetic approach predicts the most accurate solutions but typically with higher CPUs.

Table 2.5 summarizes the treatments of aerosol–cloud interactions and cloud processes. Aerosol activation by cloud droplets to form CCN is an important process affecting simulations of aerosol–cloud interactions, and aerosol direct and indirect forcing. CAM uses empirical, prescribed activated mass fraction for bulk CCN. MIRAGE and WRF/Chem use a mechanistic, parameterized activation module that is based on Köhler theory to simulate bulk CCN. Important parameters for activation such as the peak supersaturation, S_{max}, mass of activated aerosols, and the size of the smallest aerosol activated are calculated using a parametrization of Abdul-Razzak et al. (1998) and Abdul-Razzak and Ghan (2000) that relate the aerosol number activated directly to fundamental aerosol properties. GATOR/GCMOM also simulates a mechanistic, size- and composition-resolved CCN/IDN based on Köhler theory. One difference between the treatments in GATOR/GCMOM and MIRAGE is that the MIRAGE activation parametrization neglects size-dependence of the water vapor diffusivity coefficient and mass transfer coefficient, which may lead to an underestimation of cloud droplet number concentration. In addition, the equilibrium Köhler theory may be inappropriate for larger particles due to the kinetic effect (i.e., mass transfer limitation). Such size-dependence and kinetic effect are accounted for in GATOR/GCMOM. A more detailed description of US integrated models along with example case studies can be found in Zhang (2008).

2.4 Major Challenges and Future Directions

Significant progress has been made in the past two decades in the development of on-line coupled climate- (or meteorology-) chemistry and their applications for modelling global/regional climate, meteorology, and air quality, as well as the

Table 2.5 Treatments of aerosol–cloud interactions and cloud processes of on-line models

Model system	Aerosol water uptake	Aerosol activation aero-CCN/IDN	In-cloud scavenging	Below-cloud scavenging	Sedimentation of aerosols and cloud droplets
GATOR/GCMOM	Size-resolved equilibrium with RH; ZSR equation; simulated MDRH; Hysteresis is treated	Mechanistic, size- and composition-resolved CCN/IDN based on Köhler theory	Size-resolved aerosol activation, Nucl. scavenging (rainout), autoconversion for size-resolved cloud droplets or precip. rate dependent of aerosol size and composition	Size-resolved aerosol-hydrometeor coag. (washout), calculated precip. rate dependent of aerosol size and composition	Two-moment size-dependent sedimentation for all aerosols and hydrometeors
WRF/Chem	The same as MIRAGE2 but sectional (MOSAIC)	The same as MIRAGE2 but sectional (MOSAIC); bulk CNN only	similar to MIRAGE2 but sectional	similar to MIRAGE2 but sectional	The same as MIRAGE2
CAM3	For external mixtures only, equilibrium with RH, no hysteresis	Empirical, prescribed activated mass fraction; bulk CCN only	Prescribed bulk activation, autoconversion, precip. rate independent of aerosols	Prescribed bulk scav. efficiency, no-size dependence	Bulk cloud/ice sedimentation
MIRAGE2	Bulk equilibrium with RH based on Köhler theory, Hysteresis is treated	Mechanistic, parameterized activation based on Köhler theory; bulk CCN only	Modal activation, Brownian diffusion, autoconversion, precip. rate independent of aerosols	Calculated modal scaveng. coeff. using a parameterization of the collective efficiency of aerosol particles by rain drops with size dependence	Two-moment sedimentation for aerosols, nosedimentation for cloud droplets/ices

(*continued*)

Table 2.5 (continued)

Model system	Aerosol water uptake	Aerosol activation aero-CCN/IDN	In-cloud scavenging	Below-cloud scavenging	Sedimentation of aerosols and cloud droplets
Caltech unified GCM (Global)	Bulk equilibrium, ZSR equation, no hysteresis	None	Autoconversion nucl. scavenging with prescribed scavenging coefficient for sea-salt and dust and a first-order precipitation-dependent parametrization for other aerosols; precip. rate independent of aerosols	First-order precipitation-dependent bulk parametrization; calculated scavenging efficiency with size dependence	Implicitly accounted for in a parametrization of the limiting autoconversion rate

entire earth system. Several major challenges exist. First, accurately representing climate–aerosol–chemistry–cloud–radiation feedbacks in 3D climate- or meteorology-chemistry models at all scales will remain a major scientific challenge in developing a future generation of coupled models. There is a critical need for advancing the scientific understanding of key processes. Second, representing scientific complexity within the computational constraint will continue to be a technical challenge. Key issues include (1) the development of benchmark model and simulation and the use of available measurements to characterize model biases, uncertainties, and sensitivity and to develop bias-correction techniques (e.g., chemical data assimilation); (2) the optimization/parametrization of model algorithms with an acceptable accuracy. Third, integrated model evaluation and improvement and laboratory/field studies for an improved understanding of major properties/processes will also pose significant challenges, as they involve researchers from multiple disciplinaries and require a multidisciplinary and/or interdisciplinary approach. Key issues include (1) continuous operation of monitoring networks and remote sensing instrument to provide real-time data (e.g., AirNow and Satellite) for data assimilation/model evaluation and (2) the development of process-oriented models to isolate complex feedbacks among various modules/processes in on-line-coupled models. Finally, a unified modelling system that allows a single platform to operate over the full scale will represent a substantial advancement in both the science and the computational efficiency. Major challenges include globalization/downscaling with consistent model physics and two-way nesting with mass conservation and consistency. Such a unified global-to-urban scale modelling system will provide a new scientific capability for studying important problems that require a consideration of multi-scale feedbacks.

Acknowledgements The work was supported by the the US EPA-Science to Achieve Results (STAR) program (Grant # R83337601), US EPA/Office of Air Quality Planning & Standards via RTI International contract #4-321-0210288, the NSF Career Award No. Atm-0348819, and the Memorandum of Understanding between the US Environmental Protection Agency (EPA) and the US Department of Commerce's National Oceanic and Atmospheric Administration (NOAA) and under agreement number DW13921548. Thanks are due to Steve Ghan and Richard Easter at PNNL, Mark Z. Jacobson at Stanford University, and Alexander Baklanov at Danish Meteorological Institute for helpful discussions for MIRAGE/CAM3, GATOR/GCMOM, and early on-line models in Russia, respectively.

Appendix – List of Acronyms and Symbols

Acronym	Definition
3D	Three-dimensional
APC	The analytical predictor of condensation
ASTEEM	The adaptive step time-split explicit Euler method
BC	Black carbon
CAM3	The community atmospheric model v. 3

(continued)

Acronym	Definition
CB05	The 2005 version of carbon bond mechanism
CBM-EX	The Stanford University's extended carbon bond mechanism
CBM-Z	The Carbon-bond mechanism version Z
CCM	The NCAR community climate model
CCN	Cloud condensation nuclei
CFCs	Chlorofluorocarbons
CH_4	Methane
CMAQ	The EPA's community multiple air quality
CMU	Carnegie Mellon University
CO	Carbon monoxide
CO_2	Carbon dioxide
CTMs	Chemical transport models
DEMA	The iterative dynamic effective medium approximation
DMS	Dimethyl sulfide
EQUISOLV II	The EQUIlibrium SOLVer version 2
EPA	The US Environmental Protection Agency
GCM	General circulation model
GATORG	The Gas, Aerosol, TranspOrt, Radiation, and General circulation model
GATOR/GCMOM	The Gas, Aerosol, TranspOrt, Radiation, General Circulation, Mesoscale, Ocean Model
GATOR/MMTD (or GATORM)	The gas, aerosol, transport, and radiation air quality model/a mesoscale meteorological and tracer dispersion model
GChM	The PNNL global chemistry model
H_2O	Water
H_2SO_4	Sulfuric acid
IDN	Ice deposition nuclei
ISORROPIA	"Equilibrium" in Greek, refers to The ISORROPIA thermodynamic module
MADE/SORGAM	The Modal Aerosol Dynamics Model for Europe (MADE) with the secondary organic aerosol model (SORGAM)
MADRID	The model of aerosol dynamics, reaction, ionization, and dissolution
MARS-A	The model for an aerosol reacting system (MARS) –version A
MCCM (or MM5/ Chem)	The multiscale climate chemistry model
MESA	The multicomponent equilibrium solver for aerosols
MM5	The Penn State University (PSU)/NCAR mesoscale model
MIRAGE	The model for integrated research on atmospheric global exchanges
MOSAIC	The model for simulating aerosol interactions and chemistry
MOZART4	The model for ozone and related chemical tracers version 4
MSA	Methane sulfonic acid
MTEM	The multicomponent Taylor expansion method
NCAR	The National Center for Atmospheric Research
NH_4NO_3	Ammonium nitrate
$(NH_4)_2SO_4$	Ammonium sulfate
NO_3	Nitrate radical
NO_x	Nitrogen oxides
N_2O	Nitrous oxide
NOAA	The national oceanic and atmospheric administration
O_3	Ozone
OC	Organic carbon
$PM_{2.5}$	Particles with aerodynamic diameters less than or equal to 2.5 μm

(*continued*)

Acronym	Definition
PNNL	The Pacific Northwest national laboratory
Q_v	Water vapor
RACM	The regional atmospheric chemistry mechanism
RADM2	The gas-phase chemical mechanism of Regional Acid Deposition Model, version 2
RIs	Refractive indices
S(IV)	Dissolved sulfur compounds with oxidation state IV
SOA	Secondary organic aerosol
STAR	The US EPA-science to achieve results program
UCLA/GCM	The University of Los Angeles general circulation model
VOC	Volatile organic compound
WRF/Chem	The weather research forecast model with chemistry
ZSR	Zdanovskii-Stokes-Robinson

References

Abdul-Razzak H, Ghan SJ (2000) A parametrization of aerosol activation. 2. Multiple aerosol types. J Geophys Res 105:6837–6844

Abdul-Razzak H, Ghan SJ, Rivera-Carpio C (1998) A parametrization of aerosol activation: 1. Single aerosol type. J Geophys Res 103:6123–6132

Ackermann IJ, Hass H, Memmesheimer M, Ebel A, Binkowski FS, Shankar U (1998) Modal aerosol dynamics model for Europe: development and first applications. Atmos Environ 32(17):2981–2999

Austin J, Butchart N, Shine KP (1992) Possibility of an Arctic ozone hole in a doubled-CO_2 climate. Nature 360:221–225

Baklanov A, Gross A, Sørensen JH (2004) Modelling and forecasting of regional and urban air quality and microclimate. J Comput Technol 9:82–97

Baklanov A, Fay B, Kaminski J (2007) Overview of existing integrated (off-line and on-line) mesoscale systems in Europe, report of Working Group 2, April, http://www.cost728.org

Barth MC, Rasch PJ, Kiehl JT, Benkovitz CM, Schwartz SE (2000) Sulfur chemistry in the National Center for Atmospheric Research Community Climate Model: description, evaluation, features, and sensitivity to aqueous chemistry. J Geophys Res 105(1):1387–1416. doi:10.1029/1999JD900773

Cariolle D, Deque M (1986) Southern hemisphere medium-scale waves and total ozone disturbances in a spectral general circulation model. J Geophys Res 91(D10):10825–10846

Cariolle D, Lasserre-Bigorry A, Royer J-F, Geleyn J-F (1990) A general circulation model simulation of the springtime Antarctic ozone decrease and its impact on mid-latitudes. J Geophys Res 95(D2):1883–1898. doi:10.1029/89JD01644

Chuang CC, Penner JE, Taylor KE, Grossman AS, Walton JJ (1997) An assessment of the radiative effects of anthropogenic sulfate. J Geophys Res 102:3761–3778

Chung SH, Seinfeld JH (2005) Climate response of direct radiative forcing of anthropogenic black carbon. J Geophys Res 110(D1):1102. doi:10.1029/2004JD005441

Clark JHE (1970) A quasi-geostrophic model of the winter stratospheric circulation. Mon Weather Rev 98(6):443–461

Collins WD, Coauthors (2004) Description of the NCAR Community Atmosphere Model (CAM3). Tech. Rep. NCAR/TN-464_STR, National Center for Atmospheric Research, Boulder, CO, 226 pp

Collins WD, Rasch PJ, Boville BA, Hack JJ, McCaa JR, Williamson DL, Briegleb BP (2006a) The formulation and atmospheric simulation of the community atmosphere model, Version 3 (CAM3). J Climate 19:2144–2161

Collins WD, Bitz CM, Blackmon ML, Bonan GB, Bretherton CS, Carton JA, Chang P, Doney SC, Hack JJ, Henderson TB, Kiehl JT, Large WG, McKenna DS, Santer BD, Smith RD (2006b) The Community Climate System Model version 3 (CCSM3). J Climate 19:2122–2143

Côté J, Gravel S, Méthot A, Patoine A, Roch M, Staniforth A (1998) The operational CMC MRB Global Environmental Multiscale (GEM) model: Part I – design considerations and formulation. Mon Weather Rev 126:1373–1395

Cunnold D, Alyea F, Phillips N, Prinn R (1975) A three-dimensional dynamical-chemical model of atmospheric ozone. J Atmos Sci 32:170–194

Easter RC, Ghan SJ, Zhang Y, Saylor RD, Chapman EG, Laulainen NS, Abdul-Razzak H, Leung LR, Bian X, Zaveri RA (2004) MIRAGE: model description and evaluation of aerosols and trace gases. J Geophys Res 109 (D20210) doi:10.1029/2004JD004571

Eckman RS, Grose WL, Turner RE, Blackshear WT (1996) Polar ozone depletion: a three-dimensional chemical modelling study of its long-term global impact. J Geophys Res 101(D17):22977–22990. doi:10.1029/96JD02130

Fast JD, Gustafson WI Jr, Easter RC, Zaveri RA, Barnard JC, Chapman EG, Grell GA (2006) Evolution of ozone, particulates, and aerosol direct forcing in an urban area using a new fully-coupled meteorology, chemistry, and aerosol model. J Geophys Res 111:D21305. doi:10.1029/2005JD006721

Feichter J, Kjellstrom E, Rodhe H, Dentener F, Lelieveld J, Roelofs G-J (1996) Simulation of the tropospheric sulfur cycle in a global climate model. Atmos Environ 30:1693–1707

Ghan S, Easter R (2006) Impact of cloud-borne aerosol representation on aerosol direct and indirect effects. Atmos Chem Phys 6:4163–4174

Ghan SJ, Laulainen NS, Easter RC, Wagener R, Nemesure S, Chapman EG, Zhang Y, Leung LR (2001a) Evaluation of aerosol direct radiative forcing in MIRAGE. J Geophys Res 106:5295–5316

Ghan SJ, Easter RC, Hudson J, Breon F-M (2001b) Evaluation of aerosol indirect radiative forcing in MIRAGE. J Geophys Res 106:5317–5334

Ghan SJ, Easter RC, Chapman EG, Abdul-Razzak H, Zhang Y, Leung LR, Laulainen NS, Saylor RD, Zaveri RA (2001c) A physically-based estimate of radiative forcing by anthropogenic sulfate aerosol. J Geophys Res 106:5279–5293

Grell GA, Emeis S, Stockwell WR, Schoenemeyer T, Forkel R, Michalakes J, Knoche R, Seidl W (2000) Application of a multiscale, coupled MM5/chemistry model to the complex terrain of the VOTALP valley campaign. Atmos Environ 34:1435–1453

Grell GA, McKeen S, Michalakes J, Bao J-W, Trainer M, E-Y Hsie (2002) Real-time simultaneous prediction of air pollution and weather during the Houston 2000 Field Experiment, presented at the 4th conference on atmospheric chemistry: atmospheric chemistry and Texas Field Study, American Meteorological Society, Orlando, 13–17 Jan 2002

Grell GA, Peckham SE, Schmitz R, McKeen SA, Frost G, Skamarock WC, Eder B (2005) Fully coupled "online" chemistry within the WRF model. Atmos Environ 39:6957–6975

Hu X-M, Zhang Y, Jacobson MZ, Chan CK (2008) Evaluation and improvement of gas/particle mass transfer treatments for aerosol simulation and forecast. J Geophys Res 113:D11208. doi:10.1029/200759009588

Hunt BG (1969) Experiments with a stratospheric general circulation model III. Large-scale diffusion of ozone including photochemistry. Mon Weather Rev 97(4):287–306

Jacobson MZ (1994) Developing, coupling, and applying a gas, aerosol, transport, and radiation model to study urban and regional air pollution. Ph.D. Thesis, Dept. of Atmospheric Sciences, University of California, Los Angeles, 436 pp

Jacobson MZ (1995) Simulations of the rates of regeneration of the global ozone layer upon reduction or removal of ozone-destroying compounds. *EOS Suppl* Fall:1995, p F119. "Closing the hole," Geotimes Magazine (American Geological Institute), April 1996, p 9

Jacobson MZ (1997a) Development and application of a new air pollution modelling system – II. Aerosol module structure and design. Atmos Environ 31:131–144

Jacobson MZ (1997b) Development and application of a new air pollution modelling system. Part III: aerosol-phase simulations. Atmos Environ 31A:587–608
Jacobson MZ (2000) A physically-based treatment of elemental carbon optics: Implications for global direct forcing of aerosols. Geophys Res Lett 27:217–220
Jacobson MZ (2001a) Strong radiative heating due to the mixing state of black carbon in atmospheric aerosols. Nature 409:695–697
Jacobson MZ (2001b) Global direct radiative forcing due to multicomponent anthropogenic and natural aerosols. J Geophys Res 106:1551–1568
Jacobson MZ (2001c) GATOR-GCMM: a global- through urban-scale air pollution and weather forecast model 1. Model design and treatment of subgrid soil, vegetation, roads, rooftops, water, sea, ice, and snow. J Geophys Res 106:5385–5401
Jacobson MZ (2002) Control of fossil-fuel particulate black carbon plus organic matter, possibly the most effective method of slowing global warming. J Geophys Res 107(D19):4410. doi:10.1029/2001JD001376
Jacobson MZ (2004a) The short-term cooling but long-term global warming due to biomass burning. J Climate 17:2909–2926
Jacobson MZ (2004b) The climate response of fossil-fuel and biofuel soot, accounting for soot's feedback to snow and sea ice albedo and emissivity. J Geophys Res 109:D21201. doi:10.1029/2004JD004945
Jacobson MZ (2005) A solution to the problem of nonequilibrium acid/base gas-particle transfer at long time step. Aerosol Sci Technol 39:92–103
Jacobson MZ (2006) Effects of absorption by soot inclusions within clouds and precipitation on global climate. J Phys Chem A 110:6860–6873
Jacobson MZ (2007) Effects of ethanol (E85) versus gasoline vehicles on cancer and mortality in the United States. Environ Sci Technol. doi:10.1021/es062085v
Jacobson MZ, Lu R, Turco RP, Toon OB (1996) Development and application of a new air pollution modelling system. Part I: Gas-phase simulations. Atmos Environ 30B:1939–1963
Jacobson MZ, Seinfeld JH (2004) Evolution of nanoparticle size and mixing state near the point of emission. Atmos Environ 38:1839–1850
Jacobson MZ, Seinfeld JH, Carmichael GR, Streets DG (2004) The effect on photochemical smog of converting the U.S. fleet of gasoline vehicles to modern diesel vehicles. Geophys Res Lett 31:L02116. doi:10.1029/2003GL018448
Jacobson MZ, Kaufmann YJ, Rudich Y (2007) Examining feedbacks of aerosols to urban climate with a model that treats 3-D clouds with aerosol inclusions. J Geophys Res 112:D24205. doi:10.1029/2007JD008922
Kiehl JT, Hack JJ, Bonan GB, Boville BB, Williamson DL, Rasch PJ (1998) The National Center for Atmospheric Research Community Climate Model: CCM3. J Climate 11:1131–1149
Krosholm U, Baklanov A, Mahura A, Gross A, Sørensen JH, Kaas E, Chenevez J, Lindberg K (2007) ENVIROHIRLAM: on-line integrated system, presentation at the COST-728/NetFAM workshop on integrated systems of meso-meteorological and chemical transport models, Copenhagen, Denmark, 21–23 May
Liao H, Seinfeld JH (2005) Global impacts of gas-phase chemistry-aerosol interactions on direct radiative forcing by anthropogenic aerosols and ozone. J Geophys Res 110(D1):8208. doi:10.1029/2005JD005907
Liao H, Adams PJ, Chung SH, Seinfeld JH, Mickley LJ, Jacob DJ (2003) Interactions between tropospheric chemistry and aerosols in a unified general circulation model. J Geophys Res 108 (D1):4001. doi:10.1029/2001JD001260
Lohmann U, Feichter J, Penner JE, Leaitch R (2000) Indirect effect of sulfate and carbonaceous aerosols: a mechanistic treatment. J Geophys Res 105:12193–12206
Manins P (2007) Integrated modelling systems in Australia, CSIRO, presentation at the COST-728/NetFAM workshop on Integrated systems of meso-meteorological and chemical transport models, Copenhagen, Denmark, 21–23 May

Mathur R, Xiu A, Coats C, Alapaty K, Shankar U, Hanna A (1998) Development of an air quality modelling system with integrated meteorology, chemistry, and emissions. Proceedings of the Measurement of Toxic and Related Air Pollutants, AWMA, Cary, NC, September

Misenis C, Zhang Y (2010) An Examination of WRF/Chem: Physical Parameterizations, Nesting Options, and Grid Resolutions. Atmospheric Research 97:315–334

Pan Y, Hu X-M, Zhang Y (2008), Sensitivity of gaseous and aerosol predictions to gas-phase chemical mechanisms. Presentation at the 10th Conference on Atmospheric Chemistry, New Orleans, LA, 20–24 Jan

Penner JE (2003) Comment on "Control of fossil-fuel particulate black carbon and organic matter, possibly the most effective method of slowing global warming" by M. Z. Jacobson. J Geophys Res 108(D24):4771. doi:10.1029/2002JD003364

Rasch PJ, Boville BA, Brasseur GP (1995) A three-dimensional general circulation model with coupled chemistry for the middle atmosphere. J Geophys Res 100(D5):9041–9072. doi:10.1029/95JD00019

Rasch PJ, Barth MC, Kiehl JT, Schwartz SE, Benkovitz CM (2000) A description of the global sulfur cycle and its controlling processes in the National Center for Atmospheric Research Community Climate Model Version 3. J Geophys Res 105:1367–1385

Roelofs G-J, Lelieveld J (1995) Distribution and budget of O_3 in the troposphere calculated with a chemistry general circulation model. J Geophys Res 100(D10):20983–20998

Rose K, Brasseur G (1989) A three-dimensional model of chemically active trace species in the middle atmosphere during disturbed winter conditions. J Geophys Res 94(D13):16387–16403. doi:10.1029/89JD01092

Schell B, Ackermann IJ, Hass H, Binkowski FS, Ebel A (2001) Modeling the formation of secondary organic aerosol within a comprehensive air quality model system. J Geophys Res 106:28275–28293

Schlesimger ME, Mintz Y (1979) Numerical simulation of ozone production, transport and distribution with a global atmospheric general circulation model. J Atmos Sci 36:1325–1361

Stockwell WR, Kirchner F, Kuhn M, Seefeld S (1997) A new mechanism for regional atmospheric chemistry modelling. J Geophys Res 102:25847–25879

Textor C, Schulz M, Guibert S, Kinne S, Balkanski Y, Bauer S, Berntsen T, Berglen T, Boucher O, Chin M, Dentener F, Diehl T, Easter R, Feichter H, Fillmore D, Ghan S, Ginoux P, Gong S, Grini A, Hendricks J, Horrowitz L, Isaksen I, Iversen T, Kirkevåg A, Kloster S, Koch D, Kristjánsson JE, Krol M, Lauer A, Lamarque JF, Liu X, Montanaro V, Myhre G, Penner J, Pitari G, Reddy S, Seland O, Stier P, Takemura T, Tie X (2006) Analysis and quantification of the diversities of aerosol life cycles within AeroCom. Atmos Chem Phys 6:1777–1813

Tie X-X, Madronich S, Li G-H, Ying Z-M, Zhang R, Garcia AR, Lee-Taylor J, Liu Y-B (2007) Characterizations of chemical oxidants in Mexico City: a regional chemical dynamical model (WRF/Chem) study. Atmos Environ 41:1989–2008

Washington WM (1982) Documentation for the Community Climate Model (CCM), Version Ø. Tech. Rep. National Center for Atmospheric Research, Boulder, CO, 222 pp

Yarwood G, Rao S, Yocke M (2005) Updates to the carbon bond chemical mechanism: CB05. Final Report prepared for the U.S. Environmental Protection Agency, RT-04-00675, ENVIRON International Corporation, Novato, CA 94945

Zaveri RA, Peters LK (1999) A new lumped structure photochemical mechanism for large-scale applications. J Geophys Res 104:30387–30415

Zaveri RA, Easter RC, Fast JD, Peters LK (2008) Model for Simulating Aerosol Interactions and Chemistry (MOSAIC). J Geophys Res 113:D13204. doi:10.1029/2007JD008782

Zhang Y (2008) Online coupled meteorology and chemistry models: history, current status, and outlook. Atmos Chem Phys 8:2895–2932

Zhang Y, Seigneur C, Seinfeld JH, Jacobson MZ, Binkowski FS (1999) Simulation of aerosol dynamics: a comparative review of algorithms used in air quality models. Aerosol Sci Technol 31:487–514

Zhang Y, Easter RC, Ghan SJ, Abdul-Razzak H (2002) Impact of aerosol size representation on modelling aerosol-cloud interactions. J Geophys Res 107, doi:10.1029/2001JD001549

Zhang YB, Pun KV, Wu S-Y, Seigneur C, Pandis S, Jacobson M, Nenes A, Seinfeld J (2004) Development and application of the model for aerosol dynamics, reaction, ionization and dissolution (MADRID). J Geophys Res 109:D01202. doi:10.1029/2003JD003501

Zhang Y, Hu X-M, Howell GW, Sills E, Fast JD, Gustafson Jr. WI, Zaveri RA, Grell GA, Peckham SE, McKeen SA (2005a), Modelling atmospheric aerosols in WRF/CHEM. Oral presentation at the 2005 Joint WRF/MM5 User's Workshop, Boulder, CO, 27–30 June

Zhang Y, Hu X-M, Wang K, Huang J-P, Fast JD, Gustafson Jr. WI, Chu DA, Jang C (2005b) Evaluation of WRF/Chem MADRID with satellite and surface measurements: chemical and optical properties of aerosols. Oral presentation at the 2005 AGU Fall Meeting, San Francisco, CA, 5–9 Dec 2005

Zhang Y, Hu X-M, Wen X-Y, Schere KL, Jang CJ (2007) Simulating climate-chemistry-aerosol-cloud-radiation feedbacks in WRF/Chem: model Development and initial application. Oral presentation at the 6th Annual CMAS Models-3 User's Conference, Chapel Hill, NC, 1–3 Oct 2007

Zhang Y, Pan Y, Wang K, Fast JD, Grell GA (2010a) Incorporation of MADRID into WRF/Chem and Initial Application to the TexAQS-2000 Episode. J geophys Res 115:D18202. doi:10.1029/2009JD013443

Zhang Y, Wen X.-Y, Jang CJ (2010b) Simulating Climate-Chemistry-Aerosol-Cloud-Radiation Feedbacks in Continental U.S. using Online-Coupled WRF/Chem. Atmos Environ 44(29): 3568–3582

Zhang Y, Liu P, Liu X.-H, Pun B, Seigneur C, Jacobson MZ, Wang W.-X (2010c) Fine Scale Modeling of Wintertime Aerosol Mass, Number, and Size Distributions in Central California. J Geophys Res 115:D15207. doi:10.1029/2009JD012950

Chapter 3
On-Line Chemistry Within WRF: Description and Evaluation of a State-of-the-Art Multiscale Air Quality and Weather Prediction Model

Georg Grell, Jerome Fast, William I. Gustafson Jr, Steven E. Peckham, Stuart McKeen, Marc Salzmann, and Saulo Freitas

3.1 Introduction

Many of the current environmental challenges in weather, climate, and air quality involve strongly coupled systems. It is well accepted that weather is of decisive importance for air quality. It is also recognized that chemical species influence the weather by changing the atmospheric radiation budget as well as through cloud formation. However, in traditional air quality modeling procedures it is assumed that accurate air quality forecasts (and simulations) can be made even while ignoring much of the interactions between meteorological and chemical processes. This commonly used approach is termed off-line. Here we describe a modelling system – and some relevant applications – that include many of these coupled interactions, resulting in advanced research and forecast capabilities that lead to better understanding of complex interactive processes that are of great importance to regional and urban air quality, global climate change, and weather prediction. The resulting improved predictive capabilities could lead to more accurate health alerts, to a larger confidence when using the modelling system for regulatory purposes, and to better capabilities in predicting the consequences of an accidental or intentional release of hazardous materials. In Sect. 2 we give a brief outline of the model capabilities. WRF/Chem is a community model and is being developed by many different groups. Whenever developers are willing to provide their new implementations back to the community, these implementations are subjected to rigorous evaluation before being officially released to the public. Section 3 describes some of the data sets that are being used for evaluation. Finally, in Sect. 4 we describe ongoing and future development work.

G. Grell (✉)
National Oceanic and Atmospheric Administration (NOAA)/Earth System Research Laboratory (ESRL)/Cooperative Institute for Research in Environmental Sciences (CIRES), 325 Broadway, Boulder, CO 80305-3337, USA
e-mail: Georg.A.Grell@noaa.gov

A. Baklanov et al. (eds.), *Integrated Systems of Meso-Meteorological and Chemical Transport Models*, DOI 10.1007/978-3-642-13980-2_3,
© Springer-Verlag Berlin Heidelberg 2011

3.2 The On-Line Modelling Approach

In on-line modelling systems, chemistry is integrated simultaneously with meteorology, allowing feedback at each model time step both from meteorology to chemistry and from chemistry to meteorology. This technique more accurately reflects the strong coupling of meteorological and chemical processes in the atmosphere. The state-of-the-art tightly coupled modelling system described here is based on the Weather Research and Forecast (WRF) model. It is designed to be modular, and a single source code is maintained that can be configured for both research and operations. It offers numerous physics options, thus tapping into the experience of the broad modelling community. Advanced data assimilation systems are being developed and tested in tandem with the model. The model is designed to improve forecast accuracy across scales ranging from cloud to synoptic, which makes WRF particularly well suited for newly emerging Numerical Weather Prediction (NWP) applications in the non-hydrostatic regime. Meteorological details of this modelling system can be found in Skamarock et al. (2005); the details of the chemical aspects are covered in Grell et al. (2005), Fast et al. (2006), and Gustafson et al. (2007).

3.2.1 Grid-Scale Transport of Species

Although WRF has several choices for dynamic cores, the mass coordinate version of the model, called Advanced Research WRF (ARW) is described here. The prognostic equations integrated in the ARW model are cast in conservative (flux) form for conserved variables; non-conserved variables such as pressure and temperature are diagnosed from the prognostic conserved variables. In the conserved variable approach, the ARW model integrates a mass conservation equation and a scalar conservation equation of the form

$$\mu_t + \nabla \cdot (V\mu) = 0,$$
$$(\mu\phi)_t + \nabla \cdot (V\mu\phi) = 0.$$

In these equations μ is the column mass of dry air, V is the velocity (u, v, w), and ϕ is a scalar mixing ratio. These equations are discretized in a finite volume formulation, and as a result the model exactly (to machine roundoff) conserves mass and scalar mass. The discrete model transport is also consistent (the discrete scalar conservation equation collapses to the mass conservation equation when $\phi = 1$) and preserves tracer correlations (c.f. Lin and Rood (1996)). The ARW model uses a spatially 5th order evaluation of the horizontal flux divergence (advection) in the scalar conservation equation and a 3rd order evaluation of the vertical flux divergence coupled with the 3rd order Runge-Kutta time integration scheme. The time integration scheme and the advection scheme is described in Wicker and Skamarock (2002). Skamarock et al. (2005) also modified the advection to allow for positive definite transport.

3.2.2 Sub-Grid Scale Transport

Typical options for turbulent transport in the boundary layer include a level 2.5 Mellor-Yamada closure parametrization (Mellor and Yamada 1982), or a non-local approach implemented by scientists from the Yong-Sei University (YSU scheme, Hong and Pan, 1996). Transport in non-resolved convection is handled by an ensemble scheme developed by Grell and Devenyi (2002). This scheme takes time-averaged rainfall rates from any of the convective parametrizations from the meteorological model to derive the convective fluxes of tracers. This scheme also parameterizes the wet deposition of the chemical constituents.

3.2.3 Dry Deposition

The flux of trace gases and particles from the atmosphere to the surface is calculated by multiplying concentrations in the lowest model layer by the spatially and temporally varying deposition velocity, which is proportional to the sum of three characteristic resistances (aerodynamic resistance, sublayer resistance, and surface resistance). The surface resistance parametrization developed by Wesely (1989) is used. In this parametrization, the surface resistance is derived from the resistances of the surfaces of the soil and the plants. The properties of the plants are determined using land-use data and the season. The surface resistance also depends on the diffusion coefficient, the reactivity, and water solubility of the reactive trace gas.

The dry deposition of sulfate is described differently. In the case of simulations, in which aerosols are not calculated explicitly, sulfate is assumed to be present in the form of aerosol particles, and its deposition is described according to Erisman et al. (1994).

3.2.4 Photolysis Frequencies

Two options are available to calculate photolysis frequencies for the photochemical reactions of the gas-phase chemistry model. These are based on Madronich (1987) and Wild et al. (2000) and are also calculated on-line.

The profiles of the actinic flux are computed at each grid point of the model domain. To determine the absorption and scattering cross sections needed by the radiative transfer model, predicted values of temperature, ozone, and cloud liquid water content are used below the upper boundary of WRF. Above the upper boundary of WRF, fixed typical temperature and ozone profiles are used to determine the absorption and scattering cross sections. These ozone profiles are scaled with TOMS (Total Ozone Mapping Spectrometer) satellite observational data for the area and date under consideration.

The radiative transfer model in Madronich (1987) permits the proper treatment of several cloud layers, each with height-dependent liquid water contents. The extinction coefficient of cloud water β_c is parameterized as a function of the cloud water computed by the three-dimensional model based on a parametrization given by Slingo (1989). For the Madronich scheme used in WRF/Chem, the effective radius of the cloud droplets follows Jones et al. (1994). For aerosol particles, a constant extinction profile with an optical depth of 0.2 is applied.

3.2.5 Gas-Phase Chemistry: Hard-Coded Chemical Mechanisms

WRF/Chem can use two hard coded chemical gas-phase mechanisms. The first is an atmospheric chemical mechanism that was originally developed by Stockwell et al. (1990) for the Regional Acid Deposition Model, version 2 (RADM2) (Chang et al., 1989). The RADM2 mechanism is a compromise between chemical detail, accurate chemical predictions and available computer resources. It is widely used in atmospheric models to predict concentrations of oxidants and other air pollutants. For inorganic species, the RADM2 mechanism includes 14 stable species, 4 reactive intermediates, and 3 abundant stable species (oxygen, nitrogen and water). Atmospheric organic chemistry is represented by 26 stable species and 16 peroxy radicals. The RADM2 mechanism represents organic chemistry through a reactivity aggregated molecular approach (Middleton et al. 1990). Similar organic compounds are grouped together into a limited number of model groups through the use of reactivity weighting. The aggregation factors for the most emitted VOCs are given in Middleton et al. (1990). A QSSA (Quasi Steady State Approximation) method with 22 diagnosed species, which includes 3 derived from a lumped group, 3 constant, and 38 predicted species is used for the numerical solution. The rate equations for 38 predicted species are solved using a Backward Euler scheme.

The second option is the CBM-Z mechanism. The implementation of the CBM-Z (Zaveri and Peters 1999) is described in more detail in Fast et al. (2006). The CBM-Z photochemical mechanism (Zaveri and Peters 1999) contains 55 prognostic species and 134 reactions. CBM-Z uses the lumped-structure approach for condensing organic species and reactions, and is based on the widely used Carbon Bond Mechanism (CBM-IV). CBM-Z extends the original CBM-IV to include more long-lived species and their intermediates, revised inorganic chemistry, explicit treatment of lesser reactive paraffins such as methane and ethane, revised treatments of reactive paraffin, olefin, and aromatic reactions, inclusion of alkyl and acyl peroxy radical interactions and their reactions with NO_3, inclusion of longer lived organic nitrates and hydroperoxides, revised isoprene chemistry, and chemistry associated with dimethyl sulfide (DMS) emissions from oceans.

3.2.6 Gas-Phase Chemistry: The WRF/Chem/KPP Coupler

Coupled state-of-the-art meteorology/chemistry models such as WRF/Chem (Grell et al. 2005) typically include hundreds of reactions and dozens of chemical species. Solving the corresponding huge systems of ordinary differential equations requires highly efficient numerical integrators. In the case of hard-coded manually "tuned" solvers, even minor changes to the chemical mechanism, such as updating the mechanism by additional equations, often require recasting the equation system and, consequently major revisions of the code. This procedure is both extremely time consuming and error prone. In recent years, automatic code generation has become an appreciated and widely used tool to overcome these problems. The Kinetic PreProcessor (KPP, Damian et al. 2002; Sandu et al. 2003; Sandu and Sander 2006) is a computer program which reads chemical equations and reaction rates from an ASCII input file provided by the user and writes the program code necessary to perform the numerical integration. Efficiency is obtained by automatically reordering the equations in order to exploit the sparsity of the Jacobian. Recently, some of the KPP capabilities have been adapted for WRF/Chem. For this purpose, the latest KPP version (V2.1) was slightly modified (i.e. an additional switch has been implemented) to produce Fortran 90 modules which can be used in WRF/Chem without further modifications. Furthermore, a preprocessor for WRF/Chem has been developed which automatically generates the interface routines between the KPP generated modules and WRF/Chem, based on entries form the WRF/Chem registry and on KPP input files. This WRF/Chem/KPP coupler can be executed automatically during build time and considerably reduces the effort necessary to add chemical compounds and/or reactions to existing mechanisms as well as the effort necessary to add new mechanisms using KPP in WRF/Chem. At present equation files are available for various version of the RACM and RADM2 mechanisms (also RACM-MIM, Geiger et al. 2003).

3.2.7 Aerosol Modules

WRF/Chem has several options for modelling aerosols. These are based on the Modal Aerosol Dynamics Model for Europe (MADE/SORGAM) (Ackermann et al. 1998, Schell et al. 2001), and the Model for Simulating Aerosol Interactions and Chemistry (MOSAIC). MADE/SORGAM is a modification of the regional particulate model (Binkowski and Shankar 1995). Secondary organic aerosols (SOA) have been incorporated into MADE by Schell et al. (2001), by means of the Secondary Organic Aerosol Model (SORGAM). Since the different components of the modules are well documented in the above cited references, only a brief summary of the most important features shall be given here.

In MADE/SORGAM, the size distribution of the submicrometer aerosol is represented by two overlapping intervals, called modes, assuming a log-normal distribution within each mode. The conservation equations used to predict the aerosol distributions are similar to those for the gas phase species, with additional terms characterizing the aerosol dynamics and are formulated in terms of the integral moments. The most important process for the formation of secondary aerosol particles is the homogeneous nucleation in the sulfuric acid-water system. It is calculated by the method given by Kulmala et al. (1998). Aerosol growth by condensation occurs in two steps: the production of condensable material (vapor) by the reaction of chemical precursors, and the condensation and evaporation of ambient volatile species on aerosols. In MADE the Kelvin effect is neglected, allowing the calculation of the time rate of change of a moment for the continuum and free-molecular regime. The mathematical expressions of the rates and their derivation are given in Binkowski and Shankar (1995). For coagulation, in MADE it is assumed that during the process of coagulation the distributions remain log-normal. Furthermore, only the effects caused by Brownian motion are considered for the treatment of coagulation. The mathematical formulation for the coagulation process can be found in Whitby et al. (1991), Binkowski and Shankar (1995).

The inorganic chemistry system is based on MARS (Saxena et al. 1986) and its modifications by Binkowski and Shankar (1995), which calculates the chemical composition of a sulphate/nitrate/ammonium/water aerosol on equilibrium thermodynamics. Two regimes are considered depending upon the molar ratio of ammonium and sulphate. For values less than 2 the code solves a cubic for hydrogen ion molality, and if enough ammonium and liquid water are present it calculates the dissolved nitrate. For modal ionic strengths greater than 50, nitrate is assumed not to be present. For molar ratios of 2 or greater, all sulphate is assumed to be ammonium sulphate and a calculation is made for the presence of water. The Bromley methods are used for the calculation of the activity coefficients.

The organic chemistry is based on SORGAM (Schell et al. 2001). SORGAM assumes that SOA compounds interact and form a quasi-ideal solution. The gas/particle portioning of SOA compounds are parameterized according to Odum et al. (1996). Due to the lack of information all activity coefficients are assumed to be 1. SORGAM treats anthropogenic and biogenic precursors separately and is designed for the use of the RACM gas phase mechanism (Whitby et al. 1991).

The Model for Simulating Aerosol Interactions and Chemistry (MOSAIC) was added as one of the aerosol chemistry options by Fast et al. (2006). In contrast to the modal approach for the aerosol size distribution employed by MADE/SORGAM, MOSAIC employs a sectional approach, which is not restricted by physical and numerical assumptions inherent to the modal approach. The aerosol size distribution is divided into discrete size bins. Each bin is assumed to be internally-mixed so that all particles within a bin are assumed to have the same chemical composition. The number of size bins is flexible, although the default configuration has eight bins with simulations with six bins for particle diameters less than 2.5 μm and two size bins for particle diameters greater than 2.5 μm. Both mass and number are simulated for each bin. Particle growth or shrinkage resulting from uptake or loss of trace gases (H_2SO_4, HNO_3, HCl, NH_3, and eventually secondary organic species) is first calculated in a Lagrangian manner. Transfer of particles between bins is then calculated using either the two-moment approach

(Tzivion 1989), as in this study, or the moving-center approach (Jacobson 1997) with these growth rates.

Aerosols in MOSAIC are composed of sulfate, nitrate, ammonium, chloride, sodium, other (unspecified) inorganics, organic carbon (OC), elemental carbon (EC), water, and optionally, calcium, carbonate, and methane sulfonic acid. Both the actual aerosol water content and aerosol water content at 60% RH (assuming complete deliquescence) for each size bin are treated as prognostic species and are used to diagnose whether the aerosol particles are on the upper or lower hysteresis curve at relative humidities between the deliquescence and crystallization points. Assuming eight size bins, 88–112 prognostic species are required for the aerosol composition species, water, and number.

MOSAIC contains three new components designed to be numerically efficient without sacrificing accuracy including (1) a new mixing rule, called the Multicomponent Taylor Expansion Method (MTEM), to calculate the activity coefficients of various electrolytes in multi-component aqueous solutions (Zaveri et al. 2005a) (2) an efficient solid-liquid equilibrium solver, called the Multi-component Equilibrium Solver for Aerosols (MESA), to compute the solid, liquid, or mixed-phase state of aerosols (Zaveri et al. 2005b), and (3) a dynamic integration technique, called Adaptive Step Time-split Explicit Euler Method (ASTEEM), for solving the coupled gas-aerosol partitioning differential equations.

3.2.8 The Aerosol Direct Effect

A key component for on-line models is the available short-wave radiation parametrizations and how these are coupled to the various aerosol modules. In WRF/Chem, the radiation schemes include a very simple scheme (Dudhia 1989), which has aerosol effects included in a very rudimentary way (Grell et al. 2005), but also very sophisticated schemes, such as the short wave radiation package from NCAR's Community Atmospheric Model (CAM) Version 3 (http://www.ccsm.ucar.edu/models/atm-cam/index.html). Fast et al. (2006) were the first to incorporate a more sophisticated aerosol/radiation feedback into WRF/Chem. This approach is now fully implemented and tested within the WRF/Chem repository. In the current release it must be used together with the CBM-Z gasphase chemical mechanism. For release V3 (to be released in March of 2008) all other mechanisms will be available. Aerosol chemical properties and aerosol optical properties are related in the following way (Fast et al. 2006). The extinction, single-scattering albedo, and the asymmetry factor for scattering are computed as a function of wavelength and three-dimensional position. In MOSAIC, each chemical constituent of the aerosol is associated with a complex index of refraction. For each size bin, the refractive index is found by volume averaging, and Mie theory is used to find the extinction efficiency, the scattering efficiency, and the intermediate asymmetry factor as functions of the size parameter. Optical properties are then determined by summation over all size bins. Once the aerosol radiative properties are found, they are fed to a shortwave radiative transfer model to calculate the direct aerosol forcing (Fast et al. 2006).

3.2.9 The Aerosol Indirect Effect

Details of the implementation of the aerosol indirect effects can be found in Gustafson et al. (2007). In short, the first and second aerosol indirect effects are implemented in WRF/Chem through a tight coupling of the MOSAIC aerosol module to the Cloud Condensation Nulei (CCN) and cloud droplets of at least on of the microphysics and radiation schemes (Gustafson et al. 2007). Included are also parametrizations for activation/resuspension and wet scavenging (Easter et al. 2004; Ghan et al. 2001), and aqueous chemistry (Fahey and Pandis 2001). Work is in progress to add the capability to include the aerosol indirect effects into additional microphysics models that are available in WRF/Chem, as well as to the MADE/SORGAM aerosol module.

3.2.10 Fire Plumerise

A 1-dimensional (1D) fire plumerise model is included in WRF/Chem. Vegetation fires emit hot gases and particles, which are transported upward with the positive buoyancy generated by the fire. This sub-grid scale vertical transport mechanism is simulated by embedding a 1D entrainment plume model. Satellite estimates provide the intensity of the fire (in terms of size, heat fluxes and emissions) and the 1D model determines the injection height and thickness. This injection layer is used to release material emitted during the flaming phase. Details of the plumerise model can be found in Freitas et al. 2006, 2007.

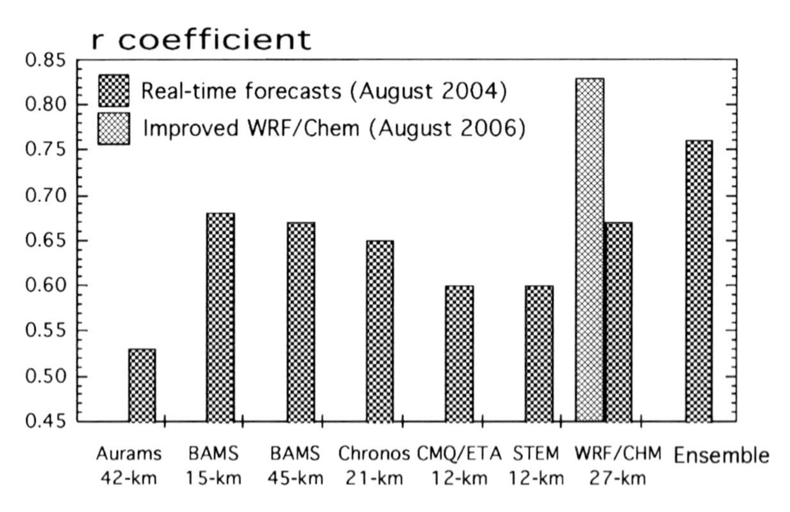

Fig. 3.1 Correlation coefficients for various model runs, comparing model forecasts of 8-h averaged peak ozone mixing ratios with those observed by surface monitoring stations. The statistics include a month worth of forecast runs

3.3 Model Evaluation and Scientific Applications

Each new implementation is subjected to rigorous evaluation, using two different test-bed data sets that include chemical constituents as well as meteorological parameters in three dimensions. These data sets are well documented and have been used for model evaluations with many different modelling systems, including WRF/Chem (McKeen et al. 2005, 2006; Pagowski et al. 2006; Pagowski and Grell 2006; Kim et al. 2006; Wilczak et al. 2006), with many more publications in preparation for the 2006 TEXAQS field experiment data set. As an example we show in Fig. 3.1 a

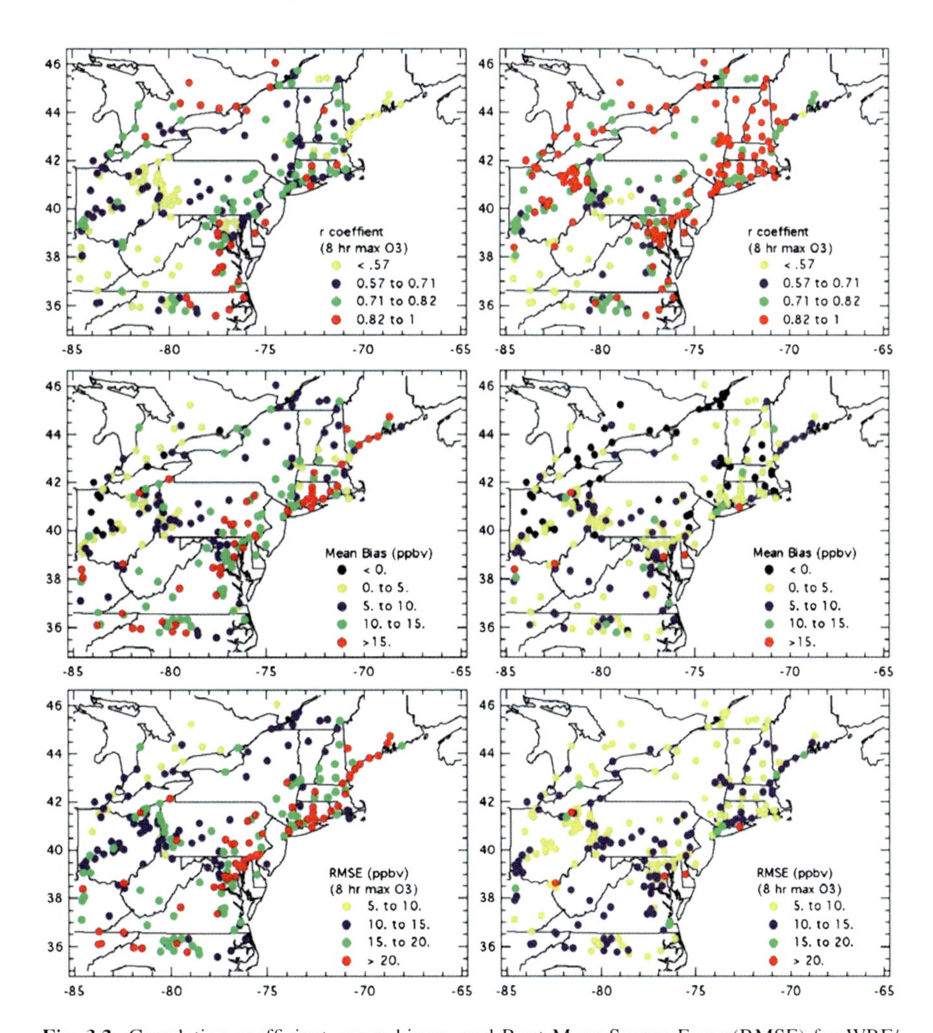

Fig. 3.2 Correlation coefficient, mean biases, and Root Mean Square Errors(RMSE) for WRF/Chem, comparing model forecasts of 8-h averaged peak ozone mixing ratios with those observed by surface monitoring stations. The statistics span a time period of 30 days. The model was run once a day at 0000UTC

comparison with observations of two particular set-ups of WRF/Chem as well as other modelling systems. The 2004 data set is based on the New England Air Quality field experiment (NEAQS2004) that took place in July and August of 2004. Displayed in Fig. 3.1 are the correlation coefficients comparing model forecasts (true forecasts, not hindcasts) of the 8-h averaged peak ozone mixing ratios when compared to surface monitoring stations. Fig. 3.2 shows the horizontal distribution of WRF/Chem results using the best performing set-up shown in Fig. 3.1). The horizontal plots are correlation coefficients, biases and root mean square errors averaged over a 30-day period (30 runs, each run is initialized at 00UTC).

Figure 3.3 and Table 3.1 are shown here to indicate the spread of available observations for model evaluation. Figure 3.3 shows a comparison of water vapor mixing ratio as predicted by WRF/Chem for the same time period, but compared to aircraft observations. Similar comparisons are available for other meteorological parameters as well as many chemical constituents, including Ozone, PM species and ozone precursors (Table 3.1). Detailed results are displayed on the WEB at http://www.al.noaa.gov/ICARTT/modeleval.

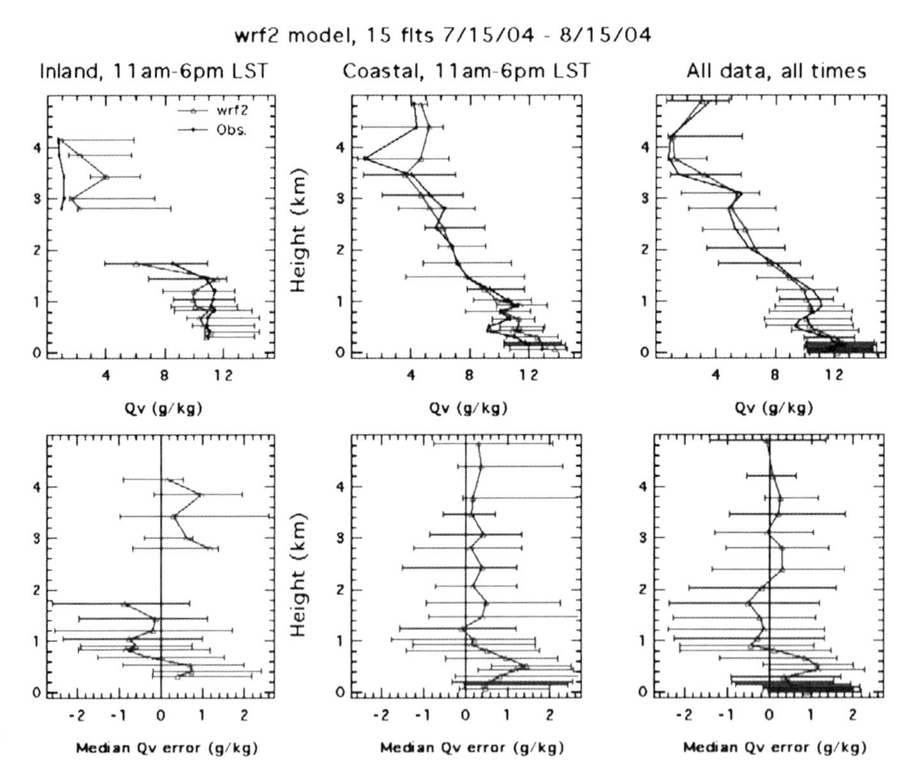

Fig. 3.3 Meteorological evaluation of WRF/Chem using the NEAQS2004 field experiment. Displayed are model predicted mean profiles of mixing ratio versus the observed profiles during 15 aircraft flights

Table 3.1 Availability of chemical constituents as well as meteorological parameters for model evaluation (from http://www.al.noaa.gov/ICARTT/modeleval)

	AURAMS	CHRONOS	STEM	WRF-2
Gas phase chemistry				
O_3	√	√	√	√
CO			√	√
NO	√	√	√	√
NO_x	√	√	√	√
NO_y	√	√	√	√
PAN	√	√	√	√
Isoprene	√	√	√	√
SO_2	√	√	√	√
NO_3	√	√		√
N_2O_5	√	√		√
CH_3CHO	√			√
Toluene		√		√
Ethylene	√	√		√
NH_3				√
Aerosols, radiation, meteorology				
PM2.5	√	√	√	√
Asol SO_4	√	√	√	√
Asol NH_4	√			√
Asol OC	√	√	√	√
Asol EC	√		√	√
Asol NO_3	√			√
JNO_2				√
T	√	√	√	√
P	√		√	√
H_2O	√	√	√	√
Winds	√	√	√	√
SST	√			√
Radiation	√	√	√	√

3.4 Ongoing Work with WRF/Chem

Ongoing and future work includes the extension of the modelling system to global scales. The global WRF model will be available with the next release in March of 2008. An off-line version exists and will soon be released for public use. Various groups are working on implementation of new aerosol modules as well as parametrizations (bin resolved and double moment bulk microphysics as well as radiation schemes) that include the aerosol direct and indirect effects. In addition, a version of the Model of Emissions of Gases and Aerosols from Nature (MEGAN) has been implemented into WRF/Chem (courtesy of Serena Chung, Jerome Fast, Christine Wiedinmyer) and will be released in March of 2008. The 2008 release will also include a new photolysis radiation scheme, known as the F-TUV scheme from Madronich. Finally, experiments using chemical data assimilation methods are underways to be able to determine a more optimal initial analysis state of the chemical and meteorological atmosphere.

References

Ackermann IJ, Hass H, Memmesheimer M, Ebel A, Binkowski FS, Shankar U (1998) Modal aerosol dynamics model for Europe: development and first applications. Atmos Environ 32(17):2981–2999

Binkowski FS, Shankar U (1995) The regional particulate matter model, 1. mode desription and preliminary results. J Geophys Res 100:26191–26209

Chang JS, Binkowski FS, Seaman NL, McHenry JN, Samson PJ, Stockwell WR, Walcek CJ, Madronich S, Middleton PB, Pleim JE, Lansford HH (1989) The regional acid deposition model and engineering model. State-of-Science/Technology, Report 4, National Acid Precipitation Assessment Program, Washington, D.C

Damian V, Sandu A, Damian M, Potra F, Carmichael GR (2002) The kinetic preprocessor KPP – a software environment for solving chemical kinetics. Comput Chem Eng 26:1567–1579

Dudhia J (1989) Numerical study of convection observed during winter monsoon experiment using a mesoscale two-dimensional model. J Atmos Sci 46:3077–3107

Easter RC, Ghan SJ, Zhang Y, Saylor RD, Chapman EG, Laulainen NS, Abdul-Razzak H, Leung LR, Bian X, Zaveri RA (2004) MIRAGE: model description and evaluation of aerosols and trace gases. J Geophys Res 109:doi: 10.1029/2004JD004571

Erisman JW, van Pul A, Wyers P (1994) Parametrization of surface resistance for the quantification of atmospheric deposition of acidifying pollutants and ozone. Atmos Environ 28: 2595–2607

Fahey KM, Pandis SN (2001) Optimizing model performance: variable size resolution in cloud chemistry modelling. Atmos Environ 35:4471–4478

Fast JD, WI Gustafson, RC Easter, RA Zaveri, JC Barnard, EG Chapman, GA Grell, SE Peckham (2006) Evolution of ozone, particulates, and aerosol direct radiative forcing in the vicinity of Houston using a fully coupled meteorology-chemistry-aerosol model. J Geophys Res-Atmos. J Geophys Res 111:D21305. doi:10.1029/2005JD006721

Freitas SR, Longo KM, Andreae MO (2006) Impact of including the plume rise of vegetation fires in numerical simulations of associated atmospheric pollutants. Geophys Res Lett 33:L17808, doi:10.1029/2006GL026608

Freitas SR, Longo KM, Chatfield R, Latham D, Silva Dias MAF, Andreae MO, Prins E, Santos JC, Gielow R, Carvalho JA Jr (2007) Including the sub-grid scale plume rise of vegetation fires in low resolution atmospheric transport models. Atmos Chem Phys 7:3385–3398

Geiger H, Barnes I, Benter T, Spitteler M (2003) The tropospheric degradation of isoprene: an updated module for the Regional Atmospheric Chemistry Mechanism. Atmos Environ 37:1503–1519

Ghan S, Laulainen N, Easter R, Wagener R, Nemesure S, Chapman E, Zhang Y (2001) Evaluation of aerosol direct radiative forcing in MIRAGE. J Geophys Res 106(D6):5295–5316

Grell GA, Devenyi D (2002) A generalized approach to parameterizing convection combining ensemble and data assimilation techniques. Geophys Res Lett 29(14):doi: 10.1029/2002GL015311

Grell GA, Peckham SE, McKeen S, Schmitz R, Frost G, Skamarock WC, Eder B (2005) Fully coupled "online" chemistry within the WRF model. Atmos Environ 39:6957–6975

Gustafson WI Jr, Chapman EG, Ghan SJ, Easter RC, Fast JD (2007) Impact on modeled cloud characteristics due to simplified treatment of uniform cloud condensation nuclei during NEAQS 2004. Geophys Res Lett 34, L19809, doi:10.1029/2007GL030021

Hong S-Y, Pan H-L (1996) Nonlocal boundary layer vertical diffusion in a medium-range forecast model. Mon Weather Rev 124:2322–2339

Jacobson MZ (1997) Development and application of a new air pollution modelling system – II. Aerosol phase simulations. Atmos Environ 31:587–608

Jones A, Roberts DL, Slingo A (1994) A climate model study of indirect radiative forcing by anthropogenic sulphate aerosols. Nature 370:450–453

Kim SW, Heckel A, McKeen SA, Frost GJ, Hsie EY, Trainer MK, Richter A, Burrows JP, Peckham SE, Grell GA (2006) Satellite-observed US power plant NOx emission reductions and their impact on air quality. Geophys Res Lett 33(L22812):5

Kulmala M, Laaksonen A, Pirjola L (1998) Parametrization for sulphuric acid/water nucleation rates. J Geophys Res 103:8301–8307

Lin S-J, Rood RB (1996) Multidimensional flux-form semi-Lagrangian transport schemes. Mon Weather Rev 124:2046–2070

Madronich S (1987) Photodissociation in the atmosphere. 1: Actinic flux and the effects of ground reflections and clouds. J Geophys Res 92:9740–9752

McKeen SA, J Wilczak, GA Grell, I Djalalova, S Peckham, E-Y Hsie, W Gong, V Bouchet, S Menard, R Moffet, J McHenry, J McQueen, Y Tang, GR Carmichael, M Pagowski, A Chan, T Dye, G Frost, P Lee, R Mathur (2005) Assessment of an ensemble of seven real-time ozone forecasts over Eastern North America during the summer of 2004. J Geophys Res 110, D21307, doi: 10.1029/2005JD005858

McKeen S, Chung SH, Wilczak J, Grell G, Djalalova I, Peckham S, Gong W, Bouchet V, Moffet R, Tang Y, Carmichael GR, Mathur R, Yu S (2006) The evaluation of several 21 PM2.5 forecast models using data collected during the ICARTT/NEAQS 2004 field 22 study. J Geophys Res 112, D10S20, doi: 2006JD007608

Mellor GL, Yamada T (1982) Development of a turbulent closure model for geophysical fluid problems. Reviews of Geophysics and Spacephysics 20:851–875

Middleton P, Stockwell WR, Carter WPL (1990) Aggregation and analysis of volatile organic compound emissions for regional modelling. Atmos Environ 24:1107–1133

Odum JR, Hoffmann T, Bowman F, Collins D, Flagan RC, Seinfeld JH (1996) Gas/particle partitioning and secondary organic aerosol yields. Environmental Science Technology 30:2580–2585

Pagowski M, Grell GA (2006) Ensemble-based ozone forecasts: skill and economic value. J Geophys Res-Atmos 111, D23S30, doi: 10.1029/2006JD007124

Pagowski M, Grell GA, Devenyi D, Peckham SE, McKeen SA, Gong W, Delle Monache L, McHenry JN, McQueen J, Lee P (2006) Application of dynamic linear regression to improve the skill of ensemble-based deterministic ozone forecasts. Atmos Environ 40:3240–3250

Sandu A, Sander R (2006) Technical note: Simulating chemical systems in fortran90 and matlab with the kinetic preprocessor KPP-2.1. Atmos Chem Phys 6:187–195

Sandu A, Daescu D, Carmichael GR (2003) Direct and adjoint sensitivity analysis of chemical kinetic systems with KPP: I – theory and software tools. Atmos Environ 37:5083–5096

Saxena P, Hudischewskyj AB, Seigneur C, Seinfeld JH (1986) A comparative study of equilibrium approaches to the chemical characterization of secondary aerosols. Atmos Environ 20:1471–1483

Schell B, Ackermann IJ, Hass H, Binkowski FS, Ebel A (2001) Modelling the formation of secondary organic aerosol within a comprehensive air quality model system. J Geophys Res 106:28275–28293

Skamarock WC, JB Klemp, J Dudhia, DO Gill, DM Barker, W Wang, JG Powers (2005) A description of the advanced research wrf version 2, Tech. Rep. 21 NCAR/TN-468+STR, NCAR

Slingo A (1989) A GCM parametrization for the shortwave radiative properties of water clouds. J Atmos Sci 46:1419–1427

Stockwell WR, Middleton P, Chang JS (1990) The second-generation regional acid deposition model chemical mechanism for regional atmospheric chemistry modelling. J Geophys Res 95:16343–16367

Tzivion S, Feingold G, Levin Z (1989) The evolution of raindrop spectra. Part II: Collisional collection/breakup and evaporation in a rainshaft. J Atmos Sci 46:3312–3327

Wesely ML (1989) Parametrization of surface resistance to gaseous dry deposition in regional numerical models. Atmos Environ 16:1293–1304

Whitby ER, McMurry PH, Shankar U, Binkowski FS (1991) Modal aerosol dynamics modelling, Rep. 600/3-91/020, Atmospheric Research and Exposure Assessment Laboratory, U.S.

Environmental Protection Agency, Research Triangle Park, NC. (Available as NTIS PB91-1617291AS from National Technical Information Service, Springfield, VA)

Wicker LJ, Skamarock WC (2002) Time-splitting methods for elastic models using forward time schemes. Mon Weather Rev 130(9):2088–2097

Wilczak J, McKeen S, Djalalova I, Grell G, Peckham S, Gong W, Bouchet V, Moffet R, McHenry J, McQueen J, Lee P, Tang Y, Carmichael GR (2006) Bias-corrected ensemble and probabilistic forecasts of surface ozone over eastern North America during the summer of 2004. J Geophys Res-Atmos 111:D23S28. doi:10.1029/2006JD007598

Wild O, Zhu X, Prather MJ (2000) Fast-J: Accurate simulation of in-and below-cloud photolysis in tropospheric chemical models. J Atmos Chem 37(3):245–282. doi:10.1023/A:1006415919030

Zaveri RA, Peters LK (1999) A new lumped structure photochemical mechanism for large-scale applications. J Geophys Res 104(D23):30387–30415

Zaveri RA, Easter RC, Peters LK (2005a) A computationally efficient multicomponent equilibrium solver for aerosols (MESA). J Geophys Res 110:D24203, doi: 10.1029/2004JD005618

Zaveri RA, Easter RC, Wexler AS (2005b) A new method for multicomponent activity coefficients of electrolytes in aqueous atmospheric particles. J Geophys Res 110(21) D02201, doi: 10.1029/2004JD004681

Chapter 4
Multiscale Atmospheric Chemistry Modelling with GEMAQ

Jacek Kaminski, Lori Neary, Joanna Struzewska, and John C. McConnell

4.1 Introduction

The strategic objective of our project was to develop and evaluate a modelling system for tropospheric chemistry and air quality (AQ). In our design we have selected the Global Environmental Multiscale model (GEM) Cote et al. (1998a) as a host meteorological model for inclusion of AQ processes. The GEM model was developed at the Canadian Meteorological Centre and is used for operational numerical weather prediction (NWP) in Canada. The GEM model was augmented by implementing AQ chemistry, including the gas phase, aerosol and cloud particles, limited wet chemistry, emission, deposition, and transport processes.

The integrated model (so called GEM-AQ) serves as a platform for performing scientific studies on processes and applications. In order to develop an AQ modelling system which can accommodate various scales and processes, we have used the GEM model as a computational platform and environmental processes were implemented on-line. There is a growing recognition for on-line implementation of tightly coupled environmental processes. Similar implementation of environmental processes is done in WRF/Chem (Weather Research and Forecasting model with Chemistry) (Grell et al. 2005), MC2-AQ (Mesoscale Compressible Community model with Air Quality) Kaminski et al. (2002), MESSy (Modular Earth Submodel System) Jockel et al. (2006), RAMS (Regional Atmospheric Modelling System) (Marcal et al. 2006) and Meso-nh (non-hydrostatic mesoscale atmospheric model) (Tulet et al. 2003).

The on-line implementation of environmental processes in the GEM model allows running in global uniform, global variable, and limited area configurations, allowing for multiscale chemical weather forecasting (CWF) modelling. This approach provides access to all required dynamics and physics fields for chemistry at every time step. The on-line implementation of chemistry and aerosol processes

J. Kaminski (✉)
Atmospheric Modelling and Data Assimilation Laboratory, Centre for Research in Earth and Space Science, York University, Toronto, Canada
e-mail: jwk@wxprime.com

A. Baklanov et al. (eds.), *Integrated Systems of Meso-Meteorological and Chemical Transport Models*, DOI 10.1007/978-3-642-13980-2_4,
© Springer-Verlag Berlin Heidelberg 2011

will allow for introducing feedback on model dynamics and physics. The use of the GEM framework permits the incorporation of chemical data assimilation techniques into the model validation and application studies in a unified fashion.

The developed modelling system can be used to plan field campaigns, interpret measurements, and provide the capacity for forecasting oxidants, particulate matter and toxics. Also, it can be used to provide guidance to evaluate exposure studies for people, animals, crops and forests, and possibly for epidemiological studies.

4.2 Methodology

4.2.1 Host Meteorological Model

The host meteorological model used for air quality studies is the Global Environmental Multiscale (GEM) model. GEM can be configured to simulate atmospheric processes over a broad range of scales, from the global scale down to the meso-gamma scale.

4.2.2 Model Dynamics

The set of non-hydrostatic Eulerean equations (with a switch to revert to the hydrostatic primitive equations) maintain the model's dynamical validity right down to the meso-gamma scales. The time discretization of the model dynamics is fully implicit, two time-level (Cote et al. 1998a, b). The spatial discretization for the adjustment step employs a staggered Arakawa C grid that is spatially offset by half a mesh length in the meridional direction with respect to that employed in previous model formulations. It is accurate to second order, whereas the interpolations for the semi-Lagrangian advection are of fourth-order accuracy, except for the trajectory estimation (Yeh et al. 2002). The vertical diffusion of momentum, heat and tracers is a fully implicit scheme based on turbulent kinetic energy (TKE). GEM version 3.1.2 was used in the current study.

4.2.3 Model Physics

The physics package consists of a comprehensive set of physical parametrization schemes (Benoit et al. 1989). Specifically, the atmospheric boundary layer (ABL) is based on a prognostic equation for TKE. The surface temperature over land surface is calculated using the force-restore method combined with a stratified surface layer. Deep convective processes are handled by a Kuo-type convective parametrization (Kuo 1974) for the resolutions that we have adopted for this study.

The infrared radiation scheme includes the effects of water vapor, carbon dioxide, ozone, and clouds. Gravity wave drag parametrization is based on a simplified linear theory for vertically propagating gravity waves generated in a statically stable flow over mesoscale orographic. GEM physics package version 4.2 was used in the current study.

4.2.4 Air Quality Modules

Air quality modules are implemented on-line in the host meteorological model. Currently, there are 37 advected and 14 non-advected gas phase species in the model. Transport of the chemically active tracers by the resolved circulation is calculated using the semi-Lagrangian advection scheme native to GEM. The vertical transfer of trace species due to subgrid-scale turbulence is parameterized using eddy diffusion calculated by the host meteorological model. Large scale deep convection in the host model depends on the resolution: in this version of GEM-AQ we use the mass flux scheme of Zhang and McFarlane (1995) for tracer species.

4.2.5 Gas Phase Chemistry

The gas-phase chemistry mechanism currently used in the GEM-AQ model is based on a modification of version two of the Acid Deposition and Oxidants Model (ADOM) Venkatram et al. (1988), derived from the condensed mechanism of Lurmann et al. (1986). The ADOM-II mechanism comprises 47 species, 98 chemical reactions and 16 photolysis reactions. In order to account for background tropospheric chemistry, 4 species (CH_3OOH, CH_3OH, CH_3O_2, and CH_3CO_3H) and 22 reactions were added. All species are solved using a mass-conserving implicit time stepping discretization, with the solution obtained using Newton's method. Heterogeneous hydrolysis of N_2O_5 is calculated using the on-line distribution of aerosol.

Although the model meteorology is calculated up to 10 hPa, the focus of the chemistry is in the troposphere where all species are transported throughout the domain. To avoid the overhead of stratospheric chemistry in this version (a combined stratospheric/tropospheric chemical scheme is currently being developed) we replaced both the ozone and NOy fields with climatology above 100 hPa after each transport time step. This ensures a reasonable upper boundary to the troposphere, while ensuring that the transport of ozone and NOy fields to the troposphere is well characterized by the model dynamics. For ozone we used the HALOE (Halogen Occultation Experiment) climatology (e.g. Hervig et al. 1993), while NOy fields are taken from the CMAM (Canadian Middle Atmosphere Model).

Photolysis rates (J values) are calculated on-line every chemical time step using the method of Landgraf and Crutzen (1998). In this method, radiative transfer calculations are done using a delta-two stream approximation for eight spectral intervals in the UV and visible applying precalculated effective absorption cross sections. This method also allows for scattering by cloud droplets and for clouds to be presented over a fraction of a grid cell. Both cloud cover and water content are provided by the host meteorological model. The J value package used was developed for MESSy (Jockel et al. 2006) and has been implemented in GEM-AQ.

4.2.6 Aerosol Package

The current version of GEM-AQ has five size-resolved aerosols types, viz. sea salt, sulphate, black carbon, organic carbon, and dust. The microphysical processes which describe formation and transformation of aerosols are calculated by a sectional aerosol module (Gong et al. 2003). The particle mass is distributed into 12 logarithmically spaced bins from 0.005 to 10.24 μm radius. This size distribution leads to an additional 60 advected tracers. The following aerosol processes are accounted for in the aerosol module: nucleation, condensation, coagulation, sedimentation and dry deposition, in-cloud oxidation of SO_2, in-cloud scavenging, and below-cloud scavenging by rain and snow.

4.2.7 Gas-Phase Removal Processes

The effects of dry deposition are included as a flux boundary condition in the vertical diffusion equation. Dry deposition velocities are calculated from a 'big leaf' multiple resistance model (Wesely 1989; Zhang et al. 2002) with aerodynamic, quasi-laminar layer, and surface resistances acting in series. The process assumes 15 land-use types and takes snow cover into account.

GEM-AQ only has a simplified aqueous phase reaction module for oxidation of SO_2 to sulphate. Thus, for the gas phase species, wet deposition processes are treated in a simplified way. Only below-cloud scavenging of gas phase species is considered in the model. The efficiency of the rainout is assumed to be proportional to the precipitation rate and a species-specific scavenging coefficient. The coefficients applied are the same as those used in the MATCH model (Multiscale Atmospheric Transport and Chemistry Model) used by the Swedish Meteorological and Hydrological Institute (SMHI) (Langner et al. 1998).

4.2.8 Emissions

The emission dataset used for global simulations was compiled using EDGAR 2.0 (Emission Database for Global Atmospheric Research) (archived in 2000, valid for 1990) and GEIA (Global Emissions Inventory Activity) global inventories. The EDGAR 2.0 dataset was chosen for its detailed information on non-methane volatile organic compound speciation. Emission data compiled for GEM-AQ includes global fields of anthropogenic emission fluxes with $1^\circ \times 1^\circ$ resolution and natural emissions with $5^\circ \times 5^\circ$ resolution. Yearly averaged anthropogenic emissions contain different industrial sectors and non-industrial activity such as burning of agricultural wastes and fuel wood, for 14 gaseous pollutants. Monthly averaged biogenic, ocean and soil emission fluxes, as well as biomass burning (forest and savannah) emissions, have been derived for nine species (seven VOC species, CO and NO_2).

In the upper troposphere/lower stratosphere (UTLS) region sources of NOx are small, from large scale convective updrafts, stratospheric sources, aircraft and lightning. We have used the monthly mean totals of lightning NOx from the GEIA inventory (scaled from 12.2 to 2 Tg/yr) and distributed them in the horizontal according to the convective cloud distribution of the model.

4.3 Model Applications

The GEM-AQ model has been run for a number of scenarios ranging from a global uniform domain (this study), global variable resolution for regional scenarios O'Neill et al. (2006), to high resolution studies (Struzewska and Kaminski 2007), global uniform long term simulations to derive a multi-year model climatology, to examine seasonal variation and regional distribution, evaluate global emissions, and provide chemical initial and boundary conditions for high resolution model simulations (Kaminski et al. 2008). GEM-AQ has also been augmented to study persistent organic pollutants (POPs) globally (Gong et al. 2007; Huang et al. 2007).

References

Benoit R, Cote J, Methot A (1989) Inclusion of a TKE boundary layer parametrization in the Canadian Regional Finite-Element Model. Mon Weather Rev 117:1726–1750

Cote J, Desmarais J-G, Gravel S, Methot A, Patoine A, Roch M, Staniforth A (1998a) The operational CMC–MRB Global Environmental Multiscale (GEM) Model. Part II: Results. Mon Weather Rev 126:1397–1418

Cote J, Gravel S, Methot A, Patoine A, Roch M, Staniforth A (1998b) The operational CMC–MRB Global Environmental Multiscale (GEM) Model. Part I: design considerations and formulation. Mon Weather Rev 126:1373–1395

Gong SL, Barrie LA, Blanchet J-P, von Salzen K, Lohmann U, Lesins G, Spacek L, Zhang LM, Girard E, Lin H, Leaitch R, Leighton H, Chylek P, Huang P (2003) Canadian Aerosol Module: a size-segregated simulation of atmospheric aerosol processes for climate and air quality models, 1. Module development. J Geophys Res 108:4007. doi:10.1029/2001JD002002

Gong SL, Huang P, Zhao TL, Sahsuvar L, Barrie LA, Kaminski JW, Li YF, Niu T (2007) GEM/POPs: a global 3-D dynamic model for semi-volatile persistent organic pollutants - 1. Model description and evaluations. Atmos Chem Phys Discuss 7:3397–3422

Grell GA, Peckham SE, Schmitz R, McKeen SA, Frost G, Skamarock WC, Eder B (2005) Fully coupled "online" chemistry within the WRF model. Atmos Environ 39:6957–6975

Hervig ME, Russell JM III, Gordley LL, Park JH, Drayson SR (1993) Observations of aerosol by the HALOE experiment onboard UARS: a preliminary validation. Geophys Res Lett 20:1291–1294

Huang P, Gong SL, Zhao TL, Neary L, Barrie LA (2007) GEM/POPs: a global 3-D dynamic model for semi-volatile persistent organic pollutants – Part 2: global transports and budgets of PCBs. Atmos Chem Phys Discuss 7:3837–3857

Jockel P, Tost H, Pozzer A, Bruhl C, Buchholz J, Ganzeveld L, Hoor P, Kerkweg A, Lawrence MG, Sander R, Steil B, Stiller G, Tanarhte M, Taraborrelli D, van Aardenne J, Lelieveld J (2006) The atmospheric chemistry general circulation model ECHAM5/MESSy1: consistent simulation of ozone from the surface to the mesosphere. Atmos Chem Phys 6:5067–5104

Kaminski JW, Plummer DA, Neary L, McConnell JC, Struzewska J, Lobocki L (2002) First application of MC2-AQ to multiscale air quality modelling over Europe. Phys Chem Earth 27:1517–1524

Kaminski JW, Neary L, Struzewska J, McConnell JC, Lupu A, Jarosz J, Toyota K, Gong SL, Côté J, Liu X, Chance K, Richter A (2008) GEM-AQ, an on-line global multiscale chemical weather modelling system: model description and evaluation of gas phase chemistry processes. Atmos Chem Phys 8:3255–3281. doi:10.5194/acp-8-3255-2008

Kuo HL (1974) Further studies on the parametrization of the influence of cumulus convection on largescale flow. J Atmos Sci 31:1232–1240

Landgraf J, Crutzen PJ (1998) An efficient method for online calculations of photolysis and heating rates. J Atmos Sci 55:863–878

Langner J, Robinson L, Persson C, Ullerstig A (1998) Validation of the operational emergency response model at the Swedish Meteorological and Hydrological Institute using data from etex and the chernobyl accident - description, test and sensitivity analysis in view of regulator applications. Atmos Environ 32:4325–4333

Lurmann FW, Lloyd AC, Atkinson R (1986) A chemical mechanism for use in long-range transport/acid deposition computer modelling. J Geophys Res 91:10905–10936

Marcal V, Rivire ED, Held G, Cautenet S, Freitas S (2006) Modelling study of the impact of deep convection on the UTLS air composition – Part I: analysis of ozone precursors. Atmos Chem Phys 6:1567–1584

O'Neill NT, Campanelli M, Lupu A, Thulasiraman S, Reid JS, Aube M, Neary L, Kaminski JW, McConnell JC (2006) Evaluation of the GEM–AQ air quality model during the Quebec smoke event of 2002: analysis of extensive and intensive optical disparities. Atmos Environ 40: 3737–3749

Struzewska J, Kaminski JW (2007) Formation and transport of photooxidants over Europe during the July 2006 heat wave – observations and GEM–AQ model simulations. Atmos Chem Phys Discuss 7:10467–10514

Tulet P, Crassier V, Solmon F, Guedalia D, Rosset R (2003) Description of the Mesoscale Nonhydrostatic Chemistry model and application to a transboundary pollution episode between northern France and southern England. J Geophys Res 108:4021, doi:10.1029/2000JD000301

Venkatram A, Karamchandani PK, Misra PK (1988) Testing a comprehensive acid deposition model. Atmos Environ 22:737–747

Wesely ML (1989) Parametrization of surface resistances to gaseous dry deposition in regional-scale numerical models. Atmos Environ 23:1293–1304

Yeh K-S, Cote J, Gravel S, Methot A, Patoine A, Roch M, Staniforth A (2002) The CMC–MRB Global Environmental Multiscale (GEM) Model. Part III: Nonhydrostatic Formulation. Mon Weather Rev 130:339–356

Zhang GJ, McFarlane NA (1995) Sensitivity of climate simulations to the parametrization of cumulus convection in the CCC-GCM. Atmos Ocean 3:407–446

Zhang L, Moran MD, Makar PA, Brook JR, Gong S (2002) Modelling gaseous dry deposition in AURAMS: a unified regional air-quality modelling system. Atmos Environ 36:537–560

Chapter 5
Status and Evaluation of Enviro-HIRLAM: Differences Between Online and Offline Models

Ulrik Korsholm, Alexander Baklanov, and Jens Havskov Sørensen

5.1 Introduction

Chemical transport models have found usage within a wide range of disciplines serving as tools for basic research, emergency preparedness, air pollution forecasting and for decision support systems. Traditionally, these models have developed independently of meteorological short range weather forecast models and are generally either Gaussian plume, Eulerian, Lagrangian or hybrid (Eulerian-Lagrangian, Gaussian-Lagrangian) models. They are forced by pre-processed output from the meteorological models (typically every 3 h) and their ability to predict the development of tracer clouds is strongly dependent on the quality of the meteorological output from the driver. This type of coupling between the meteorological driver and the chemical transport model is termed offline.

In recent years computer power has increased dramatically and short range meteorological models have reached cloud resolving scales (5 km and below). This has prompted the development of Eulerian models which integrate all the components of the chemical transport models in the meteorological driver (GATOR, WRF-CHEM, GEM-AQ, BOLCHEM, COSMO LM-ART). Hereby, the meteorological fields (wind, humidity, temperature, cloud water content, etc.) are available at each time step of the meteorological model. This type of model is termed online.

A formal definition of online and offline models may be given as: Offline models comprise separate chemical transport models forced by output from operational meteorological models, analyzed or forecasted meteorological data from archives or data sets, pre-processed meteorological data, measurements or output from diagnostic models. Online models comprise online access models in which meteorological fields are available at each time step of the meteorological model and online coupled models in which feedbacks between meteorology and tracers are also accounted for.

U. Korsholm (✉)
Danish Meteorological Institute (DMI), Lyngbyvej 100, DK-2100 Copenhagen, Denmark
e-mail: usn@dmi.dk

A. Baklanov et al. (eds.), *Integrated Systems of Meso-Meteorological and Chemical Transport Models*, DOI 10.1007/978-3-642-13980-2_5,
© Springer-Verlag Berlin Heidelberg 2011

At cloud resolving scales the meteorological models explicitly resolves more variability than corresponding statistical parameterizations may provide at coarser resolution. In order to utilize this variability for more precise transport, dispersion, deposition and transformation of pollutants online models are needed. Further advantages of online models include: meteorological and tracer fields are on the same grid, using the same physical and dynamical parameterizations thereby potentially avoiding inconsistencies, no temporal or spatial interpolation or pre-processing is necessary, all two and three dimensional meteorological fields are available at each time step of the meteorological model, thereby avoiding loss of variability in the meteorological forcing fields, it is not necessary to handle large output files from meteorological models and there is a possibility of including feedback mechanisms. Offline models, on the other hand, are more suitable for ensembles, where the meteorological fields are reused for many perturbed runs (feedbacks are neglected), and are also easier to use for adjoint modelling. They allow for usage of many different parameterizations and may employ more flexible grid structures.

Pollution concentration fields are known to contain large temporal and meso-scale variability (Anderson et al. 2003). Such variability is typically generated by meso-scale influences in the mean flow including sea breezes, development of clouds and precipitation, frontal circulations (and associated rapid changes in wind direction), urban circulations and flow over and around orographical features. The horizontal scale of such disturbances extends from a few kilometres to several hundreds of kilometres, while the time-scale ranges from less than 1 h to days. Online models have the ability to temporally and spatially resolve meso-scale disturbances and it is expected that this leads to greater accuracy in tracer distributions, especially at cloud resolving scales. The same holds true for feedbacks such as direct, indirect and semi-direct effects, due to more precise simulations of radiative fluxes and cloud development and precipitation.

The purpose of this study is to illustrate some important differences between online and offline model systems and to evaluate transport, dispersion and deposition of the online coupled meteorological and chemical transport and dispersion model Enviro-HIRLAM (High Resolution Limited Area Model), which is developed at the Danish Meteorological Institute (DMI).

5.2 Model Description

Enviro-HIRLAM is an online coupled meteorological, chemical transport and dispersion model. It is based on a previous version HIRLAM-tracer and at its core lies DMI-HIRLAM, version 6.3.7 employed for limited area short range operational weather forecasting at DMI (Chenevez et al. 2004). For a detailed description of the features in HIRLAM the reader is referred to the HIRLAM reference guide (Undèn et al. 2002).

Point sources are parameterized by assuming that the tracer distribution is uniform within the grid box containing the release site, an assumption which, depending on the spatial resolution, is fulfilled in a well mixed boundary layer. The emission is ascribed the grid point closest to the release site in the lowest model layer, corresponding to a height of approximately 30 m above the surface. In a well mixed boundary layer this will not affect the results away from the emission grid box.

For integrated atmospheric chemical transport models the requirement of consistency, monotonicity, positive definiteness and mass conservation of the numerical schemes for tracer transport is stronger than for numerical weather prediction models. To ensure the fulfilment of these requirements for chemical species and aerosols the schemes should be harmonised so that the same conservative schemes are used for meteorological and chemical quantities. Work is progressing along these lines with the implementation of the CISL (Cell Integrated Semi Lagrangian) advection scheme (Nair and Machenhauer 2002) in Enviro-HIRLAM. Several options of advection schemes for the tracers have previously been implemented (Central Difference, Semi-Lagrangian, Bott) (Bott 1989a, b); usage depends on the experiment at hand. In order to maintain large time-steps in the solution of the meteorological dynamical equations and at the same time ensure sufficient tracer mass conservation the BOTT scheme was used for the tracers while the semi-lagrangian scheme was used for meteorological quantities. This inconsistency did not affect tracer mass significantly during the runs presented here.

Dry deposition is parameterized via a resistance approach in which resistances depend on particle size and density, land-use classification and atmospheric stability (Wesely 1989; Zanetti 1990). Wet deposition is included via below cloud scavenging (washout), using a parameterization based on precipitation rates (Baklanov and Sørensen 2001) and scavenging by snow is parameterized using the scheme by Maryon and Ryall (1996). The terminal settling velocity is considered in both the laminar case, in which Stoke's law is used and the turbulent case in which a iterative procedure is employed (Näslund and Thaning 1991). For very small particles a correction for non-continuum effects is used.

The NWP-CHEM gas-phase chemistry scheme (Korsholm 2009) was developed at the DMI for usage in online models. The scheme contains 17 advected gas-phase species and 20 reactions. Photolysis reaction rates are based on Poppe et al. (1996). A modal aerosol model with three log-normal modes, developed at DMI, will be used to treat aerosol physics (Goss and Baklanov 2004; Korsholm 2009).

In the present study horizontal diffusion was switched off. Hence, the numerical diffusion arising from the Bott scheme was the only representation of sub-grid scale horizontal eddies. In the vertical a modified version of the Cuxart, Bougeault, Redelsperger (CBR)-scheme developed for HIRLAM is employed (Cuxart et al. 2000). It is based on turbulent kinetic energy, which is a prognostic variable in the model, and a stability dependent length scale formulation. The model is hydrostatic and horizontal discretization is carried out on a rotated latitude–longitude Arakawa C grid, while in the vertical a hybrid between terrain following sigma and pressure coordinates is employed with 40 levels. A non-hydrostatic version of HIRLAM

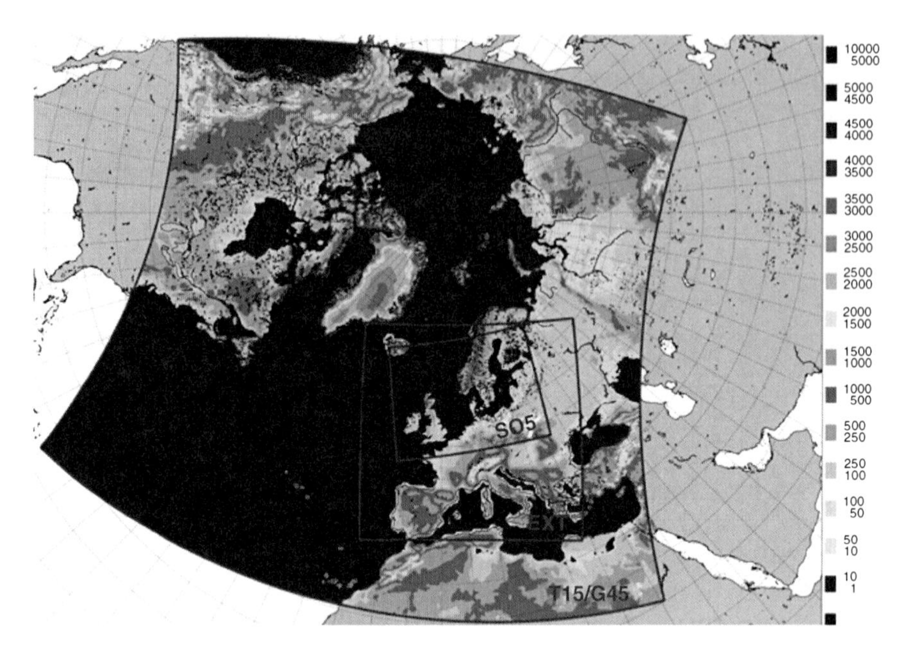

Fig. 5.1 Areas covered by the model during this study represented by surface geopotential height (meters). The domains T15 and G45 both cover the same area but the horizontal resolution is $0.15^\circ \times 0.15^\circ$ and $0.45^\circ \times 0.45^\circ$ respectively. The ETX domain is in $0.40^\circ \times 0.40^\circ$ resolution while S05 is in $0.05^\circ \times 0.05^\circ$ resolution

exists but is presently not used in Enviro-HIRLAM. Digital filter initialization is employed and the model may be run with surface and upper air (3DVAR/4DVAR) analysis. The three model areas used in this study consisted of a $0.40^\circ \times 0.40^\circ$ domain termed ETX and a $0.15^\circ \times 0.15^\circ$ domain termed T15 (Fig. 5.1).

5.3 Model Evaluation

5.3.1 Transport and Dispersion

During the first European Tracer Experiment (ETEX-1) a non-depositing tracer gas (Perflouro-Methyl-Cyclo-Hexane) was emitted from a site in Northern France (Brittany (2°00′30″, 48°03′30″)). The average emission rate was 7.95 g s^{-1} and it commenced on 23 October at 16:00 UTC lasting for 11 h and 50 min. The spatial and temporal development of the tracer cloud was measured at 168 measurement stations in Europe and both real time and retrospective model inter-comparison projects were carried out (Graziani et al. 1998; Mosca et al. 1998). The purpose of this experiment was to evaluate the models ability to transport and disperse a tracer.

Transport and dispersion was evaluated without any form of tuning by comparing a simulation of the ETEX-1 release to the official measurements of surface concentration. To facilitate comparisons with models evaluated during ATMES II (Atmospheric Transport Model Evaluation Study) an identical statistical methodology was employed (Mosca et al. 1998). Background values were subtracted so that only the pure tracer concentration was used. Measurements of zero concentration (concentrations below the background level) were included in time series to the extent that they lay between two non-zero measurements or within two before or two after a non-zero measurement. Hereby, spurious correlations between predicted and measured zero-values far away from the plume track are reduced.

The current version of Enviro-HIRLAM has not previously been evaluated against ETEX-1 measurements. The ETX domain (Fig. 5.1) was used with at time-step fixed at 10 min, and initial and boundary conditions were post-processed from the European Centre for Medium-Range Weather Forecasts operational model, IFS (Integrated Forecast System). No surface or upper air data assimilation was employed and the model was integrated 80 h into the future. The start time was on 23 October 1994 at 12:00 UTC, 4 h before the start of the release. Output was interpolated to measurement stations in order to compare to the observations and produce statistical measures.

5.3.1.1 Results and Discussion

The synoptic situation in the days following the ETEX-1 release has previously been described in detail by Gryning et al (1998) and Graziani et al. (1998). Correspondingly, the model plume was initially advected by a westerly flow, mainly influenced by synoptic-scale forcings, in a north-easterly direction (Nastrom and Pace 1998). The spatial structure of the model plume resembled the observations (Graziani et al. 1998) and remained continuous throughout the forecast period (Fig. 5.2). The plume is most sensitive to meso-scale perturbations during its initial development. Even though the bulk of the plume remains continuous the marginal structure may be affected by such disturbances and cause large errors in verification scores at specific stations. After 36 h the model plume had attained a U-shaped

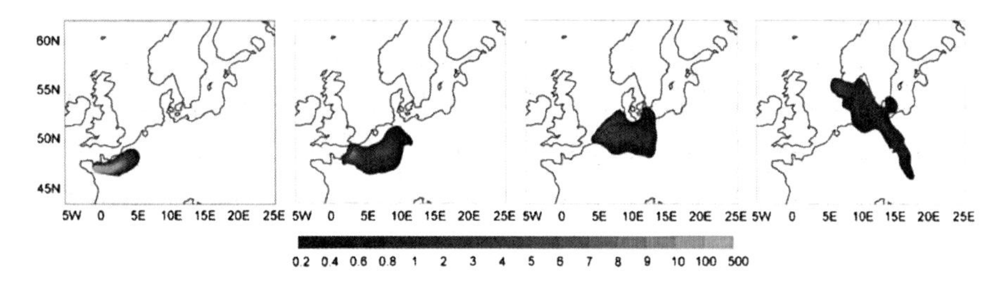

Fig. 5.2 Simulated development of the ETEX-1 tracer plume at 12, 24, 36 and 48 h after start of release (ngm^{-3}), corresponding measurements can be found in Graziani et al. (1998)

deformation receding over Northern Germany. Although less distinct a similar structure is present in the observations. The model, thus, over-predicted the development of the deformation, which extended further to the North.

In line with the measurements the model plume stretched and its axis tilted, so it was oriented in a North-West to South-East direction, after 48 h. The peak concentration, however, was located too far to the North. After 60 h the largest concentration values were found in the North Sea, a feature which is also present in the observations. Following the methodology of ATMES II the time development of the model plume was evaluated at 11 selected stations (Mosca et al. 1998). These were chosen to constitute two arcs at different distances from the release site. The first arc (measurement stations: NL05, B05, NL01, D44) follows the Eastern border of Belgium. The arrival time at these stations ranged from 15 to 18 h. The second arc (measurement stations: DK05, DK02, D42, D05, PL03, CR03, H02) extended from Denmark in the North to Hungary in the South and the arrival times at these stations ranged from 30 to 39 h. The average correlation, normalised mean square error (NMSE), bias and figure of merit in time (FMT) (Table 5.1) at the stations are 0.49, 5.25, 0.18 ng m^{-3} and 29.35 % respectively. These values are all acceptable when compared to the model scores during ATMES II. All the scores are degraded by the values at the stations in the first arc, suggesting worse performance close to the release site than further away from it, which was generally also found during ATMES II.

5.3.2 *Deposition*

To evaluate the deposition routines a simulation of the Chernobyl accident was carried out and compared to measurements of total deposited Cesium 137 (Cs-137). The measurements were extracted from the Radioactivity Environmental Monitoring database at the Joint Research Centre, Ispra, Italy (http://rem.jrc.cec.eu.int/). The comparison date was chosen to be 1 May 1986 at 12:00 UTC, since at this time the greatest number of measurements was available. Statistical measures were calculated following the recommendations of the Atmospheric Transport Model Evaluation Study (ATMES) final report (Klug et al. 1992).

The total amount and corresponding temporal development of the Cs-137 emission has been estimated (Devell et al. 1995; De Cort et al. 1998; Persson et al. 1986) and is associated with at least 50 % uncertainty. The current simulation considered

Table 5.1 Statistical scores for the ETEX-1 simulation at 11 selected measurement stations

Station	B05	CR03	D05	D44	DK02	DK05	H02	D42	NL01	NL05	PL03
Bias (ngm^{-3})	0.76	−0.08	0.02	0.45	−0.01	−0.11	−0.02	−0.14	0.48	0.65	−0.06
NMSE	12.9	7.95	2.00	4.54	0.93	4.77	1.05	2.25	4.46	14.8	1.95
Correlation	0.80	0.92	0.29	0.64	0.68	0.08	0.86	0.46	−0.05	0.29	0.43
FMT (%)	12.9	26.1	29.6	32.1	51.4	15.4	49.3	32.7	15.9	19.1	38.4

the transport, dispersion and deposition of Cs-137 and employed vertically stratified point sources in order to simulate the explosions and the following fire.

The size distribution of particles containing Cs-137 is not known and here only mono-disperse particles with a radius of 0.5 μm and a density of 1.88 g cm^{-3} were considered. The model area corresponded to the G45 domain (Fig. 5.1). The start time was at 25 April 1986 at 18 UTC and the model was run 2 days ahead and then reinitialized and restarted until 7 May at 18:00 UTC. Surface analysis and 3DVAR upper air analysis was used as initial conditions for the meteorology at the beginning of each cycle and 6 hourly boundaries were post-processed from the IFS model.

5.3.2.1 Results and Discussion

Considering the large uncertainty of the emission data, the uncertainty in the measurements, the mono-disperse nature of the simulation and the coarse resolution the model reproduce (spatially) most features of the deposition field satisfactory (Fig. 5.3a, b); De Cort et al. 1998). This includes the peaks close to the accident site,

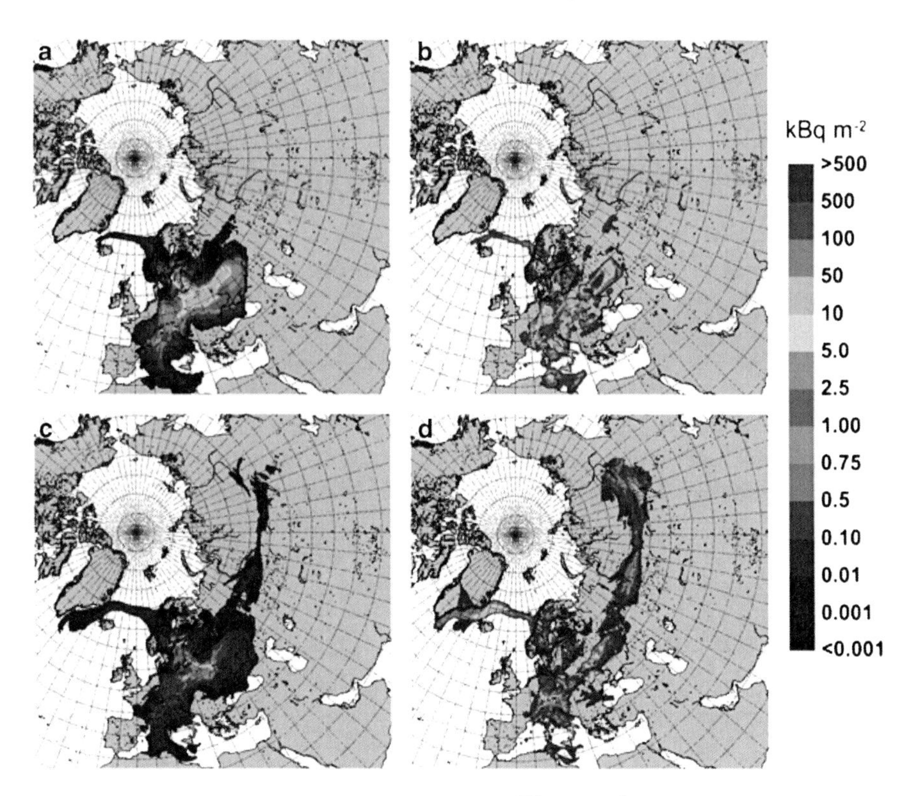

Fig. 5.3 Accumulate dry (**a**) and wet (**b**) deposited Cs^{137} ($kBqm^{-2}$) on 12 UTC 1 May 1986. (**a**) and (**b**) are predictions using 0.45° resolution while (**c**) and (**d**) are the predictions using 0.15° resolution. Corresponding measurements may be found in De Cort et al. (1998)

Table 5.2 Global (containing both temporal and spatial variability) statistical scores for the simulation of Cs-137 deposition after the Chernobyl accident for 0.45° and 0.15° resolution

Global statistical parameter	Calculated value	
	0.45°	0.15°
Observed mean ($kBqm^{-2}$)	19.97	
Predicted mean ($kBqm^{-2}$)	56.74	
NMSE	6.34	0.83
Bias ($kBqm^{-2}$)	38.77	2.17
Pearson's correlation	0.59	0.38
FMT (%)	26.29	45.11

in southern Finland, Switzerland, Austria and Italia. The band of increased activity extending from Southern Finland across Sweden and Norway is not well captured and is known to be caused by wet deposition. In the model the precipitation falls in a band further south causing the shift in the wet deposition pattern (Fig. 5.3).

The simulation is associated with large global bias (Table 5.2) which may be due to the misplacement of this band and the rest of the statistical scores are satisfactory considering that no attempts of tuning the model has been performed. Another reason for the large global bias could be insufficient horizontal resolution. The simulations were repeated in 0.15° resolution, keeping everything else as described above. The cumulated deposition fields are displayed in Fig. 5.3c, d. The band of wet deposited Cesium across Sweden and Norway is not better resolved in higher resolution. The model rain out in the Bay of Finland, instead of in Sweden, generating a large peak of deposited Cesium. Although the concentration in the emission grid box generally increases, with increasing resolution, the dry deposition field generally decreases over much of the domain. This is confirmed by the statistical scores which all improve, except for the correlation. As resolution increased so did the vertical mixing and more mass was transpoted into the jet stream reciding at about 600 hPa. The stream transported Cesium towards the east with a speed of approximately 30 m s^{-1}, leading to a band of increased dry and wet deposition in northern Russia. The increased mass at higher levels caused the surface concentration to decrease and wet deposition near the east coast of Greenland to increase. Hence, the bias improved from increasing the resolution, however, it is seen that the correlation decrease due to overestimation of the total deposited field.

5.4 On-Line/Off-Line Comparison

5.4.1 Variability

Offline models may not have the ability to temporally resolve the evolution of mesoscale disturbances which often have time scales well below the coupling interval (the time span between updates of meteorological fields for the offline model) even when sophisticated interpolation and diffusion procedures are employed to produce intermediate time steps. This may lead to errors in the simulation of tracer transport, dispersion, deposition, transformations and chemistry. The purpose of this experiment is to illustrate this difference between online and offline models.

For the current experiment a reference ETEX-1 simulation was run in online mode (using a 10 min time step) while an identical set of experiments were run in offline mode using different coupling intervals of 30, 60, 120, 240 and 360 min. The results at a particular station, which is known to be influenced by meso-scale activity, were compared to measurements and conclusions regarding the effect of the coupling interval were drawn.

The set-up is identical to what is described in the evaluation of transport and dispersion section. The output has been interpolated to two ETEX-1 measurement stations, F15 and DK02, which were dominated by short and long range transport respectively.

5.4.1.1 Results and Discussion

At station F15 the measurements are dominated by a single peak which was captured well by the simulation (Fig. 5.4; run with a 10 min coupling interval). The result was not sensitive to variations in the coupling interval of up to 6 h. This suggests that the peak was generated by transport of the bulk of the plume over the site without any influence from short time scale disturbances (less than 6 h), i.e the wind did not vary rapidly. At station DK02 the plume had traversed a region in which meso-scale disturbances are known to influence the dispersion (Sørensen et al. 1998). The predicted development of the concentration field had a phase error of a few hours on the arrival of the plume but was otherwise in good agreement with the observations (Fig. 5.4). A false (not in observations) peak preceding the plume existed and is indicative of meso-scale influences during plume development in the model. Notice that the first peak did not contribute to the statistical scores because the observations are zero. As the coupling interval was increased the main (second)

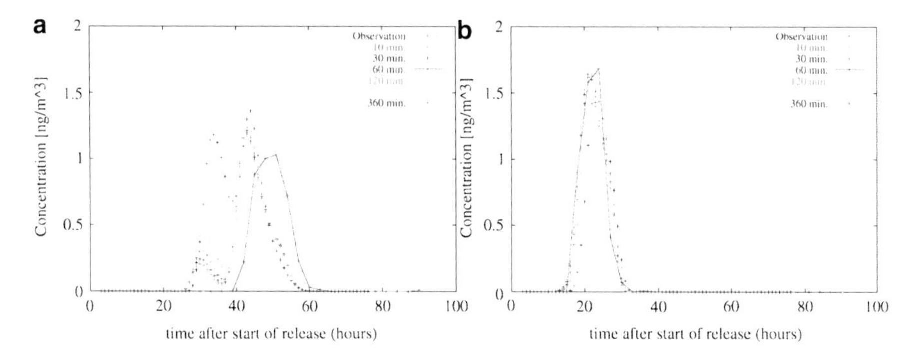

Fig. 5.4 Measured and modelled time development of concentration (ngm^{-3}) at ETEX stations DK02 (**a**) and F15 (**b**) for coupling intervals 10 (online), 30, 60, 120, 240 and 360 min

peak remained unaffected while the amplitude of the first peak gradually increased. This suggests that the existence of the first peak is related to short time-scale disturbances in the forecasted meteorological fields, while the second peak is generated by transport of the bulk of the plume.

Meso-scale eddies are superposed on the mean flow generating cyclonic and anti-cyclonic perturbations in the plume. The eddies are visible as peaks in plots of relative vorticity (Fig. 5.5; the eddies which influenced the plume development is marked with arrows). The eddies filled the boundary layer between the surface and at least 800 Hpa, persisted at least 15 h, had maxima of $\pm 6 \times 10^{-5}\,s^{-1}$ respectively at the surface and tilted towards the NE and SW with height.

At 24 h after the start of the release the plume maximum had split into two separate parts (Fig. 5.2). The head received cyclonic rotational momentum from a meso-scale disturbance and reached DK02 after 26 h giving rise to the first peak. As the latter part of the cloud progressed it received anti-cyclonic momentum and after 36 h the plume attained a U-shaped deformation which was advected towards DK02. The rotational time scale of the eddies was not large enough (compared to the advective time scale of the plume) to cause a full revolution in the plume.

As the coupling interval was increased changes in the magnitude of the eddies were not resolved and the U-shape extended further northwards leading to increased peak values in the concentration field at DK02 (Fig. 5.6). Increased temporal resolution constrained the evolution of the meso-scale disturbances, leading to better correspondence with measurements. Hence, even at coarse resolution it

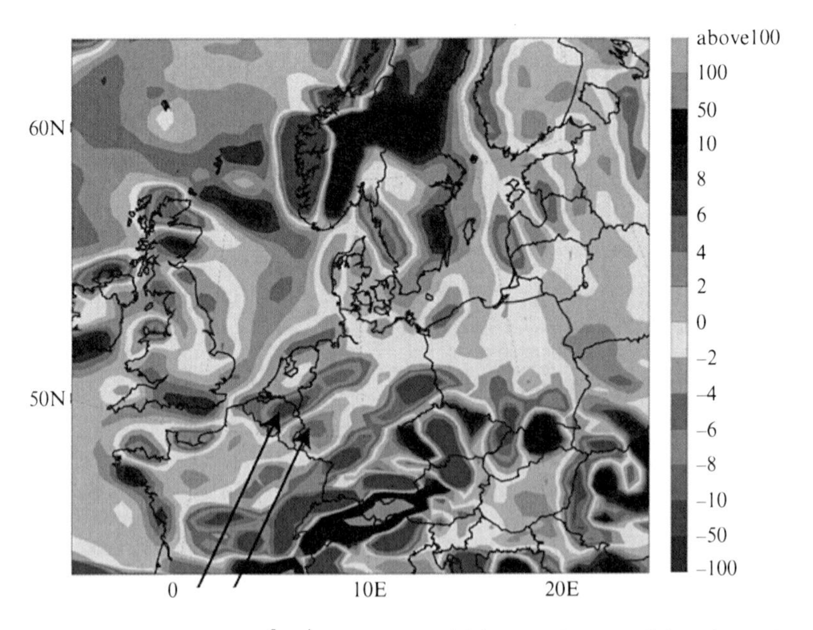

Fig. 5.5 Relative vorticity ($\times 10^5\,s^{-1}$) in lowest model layer at the start of the release. *Arrows* indicate eddies influencing the plume during transport

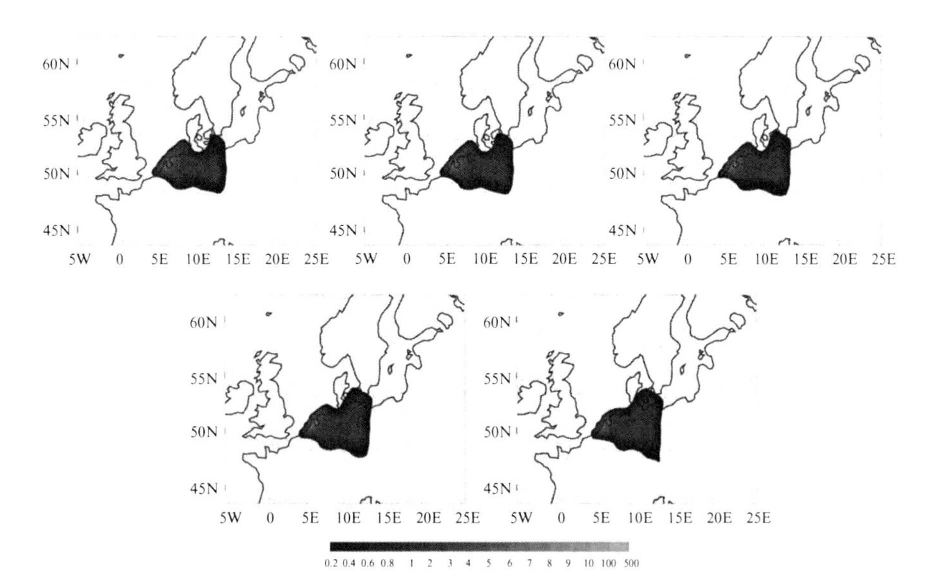

Fig. 5.6 Modelled concentration field (ngm^{-3}) 36 h after start of release for a 30, 60, 120, 240 and 360 min coupling interval respectively

may be necessary to decrease the coupling interval in order to achieve correspondence with measurements at specific stations.

This experiment was conducted in a meteorological situation without strong dispersion and only horizontal effects were considered. However, previous studies have shown that very short coupling intervals are necessary to constrain vertical mixing processes (Grell et al. 2004). In general, the appropriate length of the coupling interval will depend on the application and the meso-scale activity. From these experiments it is not possible to give general recommendations along these lines, however, a coupling interval of 3 h is not sufficient to constrain the development of the meso-scale disturbances.

5.5 Feedbacks

The presence of aerosols and trace gases in the atmosphere may affect meteorological fields through changes in cloud processes and the radiation balance. Such changes in the meteorological fields may feed back upon the aerosol and trace gas fields either directly through dynamical modifications in transport and dispersion or through changes in chemical reaction rates as temperature and cloud cover is modified. Enviro-HIRLAM contains representations of the first and second aerosol indirect effects (1IE and 2IE). As the activated aerosol fraction increases cloud droplet effective radius decrease and number concentration increase. Hereby, cloud reflectance increases and precipitation development and thereby cloud lifetime is affected.

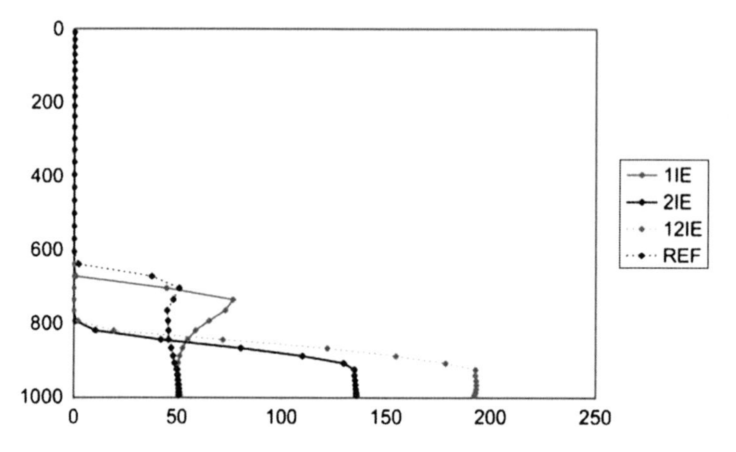

Fig. 5.7 Vertical profile of NO_2 at 12 UTC on 19 June 2005 for different runs

In a particular experiment runs with and without 1IE and 2IE were compared. The accumulation mode aerosol was boosted by a factor of about 2,000 and used as activated cloud droplets in order to test the largest possible effect on effective radius (with the purpose of investigating the parameterizations of 1IE and 2IE). By perturbing the natural background cloud droplet number concentration the effective radius and thereby the cloud radiative properties and the auto conversion term in the convection scheme (Sass 2002) were modified (in this test the effective radius decreased to the cuf-off value of 4 μm at peak cloud droplet number concentration). Model runs were done in 0.05º resolution on a domain covering 665 × 445 km centred around Paris. The meteorological case contained deep convection and both convective and stratiform precipitation. Four model runs were performed. The baseline run, denoted REF, did not contain any interactions between aerosols and meteorology, run 1IE contained the first aerosol indirect effect, run 2IE contained the second aerosol indirect effect and run 12IE contained both the first and second aerosol indirect effect. Figure 5.7 shows the vertical profile of NO_2 at 12 UTC on 19 June 2005 for the four runs at the point of maximum difference in NO_2 concentration between REF and 12IE. 2IE had a strong influence on NO_2 in the boundary layer while 1IE had some influence in the upper part. The increased cloud lifetime due to 2IE acted to increase cloud cover and thereby cool the surface during daytime. Surface temperature changed several degrees which in turn activated convective cells. As boundary layer height decreased NO_2 was strongly redistributed.

5.6 Conclusions

The online coupled model system, Enviro-HIRLAM, which integrates a meteorological model and an atmospheric chemical transport model has been described and evaluated.

- Enviro-HIRLAM performs satisfactory with regards to transport and dispersion when simulating the ETEX-1 controlled release.
- Enviro-HIRLAM performs satisfactorily with regards to deposition when simulating the Chernobyl accident.
- In situations with meso-scale activity the coupling interval, in offline models, is important in constraining meso-scale influences on plume development. Online coupling improves the results.
- In a particular test of the parameterizations of the aerosol indirect effects the NO_2 concentration was strongly affected by dynamical feedbacks associated with the aerosol indirect effects.

Acknowledgements This work was supported by the Danish Meteorological Institute and the Copenhagen Global Change Initiative (COGCI).

References

Anderson LT, Charlson JR, Winker MD, Ogren AJ, Holmen K (2003) Mesoscale variations of tropospheric aerosols. J Atmos Sci 60:119–136

Baklanov A, Sørensen JH (2001) Parameterisation of radionuclide deposition in atmospheric long-range transport modelling. Phys Chem Earth (B) 26(10):787–799

Bott A (1989a) A positive definite advection scheme obtained by non-linear renormalization of the advective fluxes. Mon Weather Rev 117:1006–1015

Bott A (1989b) Reply. Mon Weather Rev 117:2633–2636

Chenevez J, Baklanov A, Sørensen JH (2004) Pollutant transport schemes integrated in a numerical weather prediction model: model description and verification results. Meteorol Appl 11: 265–275

Cuxart J, Bougeaults P, Redelsberger JL (2000) A turbulence scheme allowing for mesoscale and large-eddy simulations. Q J Roy Meteorol Soc 126:1–30

De Cort M, Dubois G, Fridman Sh D, Germenchuk MG, Izrael Yu A, Janssens A, Jones AR, Kelly GN, Kvasnikova EV, Matveenko II, Nazarov IM, Pokumeiko Yu M, Sitak VA, Stukin ED, Tabachny L Ya, Tsaturov Yu S (1998) Atlas of Caesium Deposition on Europe after the Chernobyl Accident, EUR report nr. 16733, Office for Official Publications of the European Communities, Luxembourg, Plate 1

Devell L, Guntay S, Powers DA (1995) The Chernobyl reactor accident source term: development of a consensus view. Paris CSNI report, OECD/NEA

Goss A, Baklanov A (2004) Modelling the influence of dimethyl sulphide on the aerosol production in the maribe boundary layer. Int J Environ Pollut 22:51–71

Graziani G, Klug W, Mosca S (1998) Real-time long-range dispersion model evaluation of the ETEX first release, EUR 17754 EN. Office for official publications of the European Communities, Luxembourg. ISBN 92-828-3657-6

Grell AG, Knoche R, Peckham ES, McKeen AS (2004) Online versus offline air quality modeling on cloud-resolving scales. Geophys Res Lett 31:L16117

Gryning ES, Batchvarova E, Schneiter D, Bessemoulin P, Berger H (1998) Release site during two tracer experiments. Atmos Environ 32(24):4123–4137

Klug W, Graziani G, Grippa G, Pierce D, Tassone C (1992) Evaluation of long range atmospheric transport models using environmental radioactivity data from the Chernonyl accident, The ATMES report, Elsevier Applied Science, London

Korsholm SU (2009) Integrated modelling of aerosol indirect effects. Development and application of an online coupled chemical weather model. PhD thesis, University of Copenhagen

Maryon RH, Ryall DB (1996) Developments to the UK nuclear accident response model (NAME). Department of Environment, UK Met. Office. DOE report # DOE/RAS/96.011

Mosca S, Bianconi R, Graziani G, Klug W (1998) ATMES II: evaluation of long-range dispersion models using data of the 1st ETEX release EUR 17756 EN. Office for official publications of the European Communities, Luxembourg. ISBN 92-828-3657-X

Nair R, Machenhauer B (2002) The mass-conservative cell-integrated semi-Lagrangian advection scheme on the sphere. Mon Weather Rev 130(3):649–667

Näslund E, Thaning L (1991) On the settling velocity in a nonstationary atmosphere. Aerosol Sci Technol 14:247–256

Nastrom JS, Pace JC (1998) Evaluation of the effect of meteorological data resolution on lagrangian particle dispersion simulations using the etex experiment. Atmos Environ 32(24):4187–4194

Persson C, Rodhe H, De Geer LE (1986) The Chernobyl accident – a meteorological analysis of how radionuclides reached Sweden, SMHI/RM Report No. 55

Poppe D, Andersson-Skold Y, Baart A, Builtjes PJH et al (1996) Gas-phase reactions in atmospheric chemistry and transport models: a model intercomparison. EUROTRAC a EUREKA environmental project. EUROTRAC international scientific secretariat, Garmisch-Partenkirchen

Sass B (2002) A research version of the STRACO cloud scheme. Danish Meteorological Institute, Technical Report no. 02-10

Sørensen JH, Rasmussen A, Ellermann T, Lyck E (1998) Mesoscale influence on long-range transport -evidence from ETEX modelling and observations. Atmos Environ 32(24): 4207–4217

Undèn P, Rontu L, Järvinen H, Lynch P, Calvo J, Cats G, Cuhart J, Eerola K et al (2002) HIRLAM-5 scientific documentation. December 2002, HIRLAM-5 project report, SMHI. Norrköping, Sweden

Wesely ML (1989) Parameterisation of surface resistances to gaseous dry deposition in regional scale numerical models. Atmos Environ 23(6):1293–1304

Zanetti P (1990) Air pollution modelling – theories, computational methods and available software. Computational Mechanics/Van Nostrand Reinhold, Southampton/New York

Chapter 6
COSMO-ART: Aerosols and Reactive Trace Gases Within the COSMO Model

Heike Vogel, D. Bäumer, M. Bangert, K. Lundgren, R. Rinke, and T. Stanelle

6.1 Introduction

Atmospheric aerosol particles modify the radiative transfer in the atmosphere and they have an impact on the cloud formation. Therefore, they alter the weather and they have an impact on climate. The anthropogenic part of this modification of the state of the atmosphere is currently not well understood and it raises the largest uncertainties with respect to climate change (see the IPCC report 2007). We developed a new on-line model system to investigate the aerosol–radiation-interaction on the regional scale.

6.2 Method

Based on the mesoscale model system KAMM/DRAIS/MADEsoot/dust (Riemer et al. 2003; Vogel et al. 2006a, b) we developed an enhanced model system to simulate spatial and temporal distribution of reactive gaseous and particulate matter. The meteorological driver of the old model system (KAMM) was replaced by the operational weather forecast model COSMO model (= former *L*okal *M*odell, Steppeler et al. 2003) of the German Weather Service (DWD). The name of the new model system is COSMO-ART (ART stands for *A*erosols and *R*eactive *T*race gases; Vogel et al. 2006a, b). The atmospheric chemistry transport model (ACTM) module was on-line coupled with the operational version of the COSMO model. That means that in addition to the transport of a non-reactive tracer the dispersion of chemical reactive species and aerosols can be calculated. Secondary aerosols which are formed from the gas phase, directly emitted components like soot, mineral dust,

H. Vogel (✉)
Institut für Meteorologie und Klimaforschung, Karlsruhe Institute of Technology (KIT), Postfach 3640, 76021 Karlsruhe, Germany
e-mail: heike.vogel@kit.edu

A. Baklanov et al. (eds.), *Integrated Systems of Meso-Meteorological and Chemical Transport Models*, DOI 10.1007/978-3-642-13980-2_6,
© Springer-Verlag Berlin Heidelberg 2011

sea salt, and biological material are represented by log normal distributions. Processes such as coagulation, condensation, and sedimentation are also taken into account. The emissions of biogenic VOCs, dust particles, sea salt, and pollen are calculated also on-line, taking into account the dependencies on the meteorological variables. To calculate efficiently the photolysis frequencies a new method was developed using the GRAALS radiation scheme (Ritter and Geleyn 1992) which is already implemented in LM. With this model system we want to quantify feedback processes between aerosols and the state of the atmosphere and the interaction between trace gases and aerosols on the regional scale. To enable fully coupled model runs, the aerosol optical properties have been parameterized (Bäumer et al. 2004) since on-line Mie computations were too time-consuming.

In the parametrization that is based on off-line Mie calculations, the aerosol optical properties for the eight spectral bands of the LM radiation scheme are calculated separately for the five modes of the aerosol model as a function of dry mass density, water content and soot content in each mode. For the simulations, the climatologically aerosol optical properties, which are used in the standard LM version, are replaced by these parameterized ones that take into account current modal aerosol mass densities. By comparing different simulation results obtained with parameterized and climatological aerosol optical properties, the impact of the aerosol can be quantified not only on the radiation, but also on other meteorological variables such as temperature. The model system can be embedded by one way nesting into individual global scale models as the GME model (Global model of the DWD) or the ECMWF model. Figure 6.1 gives an overview of new model system.

6.3 Results

In the following two case studies where the model was used to quantify the impact of natural and anthropogenic aerosol particles on the regional weather will be explained.

6.3.1 The Interaction of Mineral Dust with Radiation

The first application is the simulation of a mineral dust event over West Africa in March 2004. During this event there were high wind speeds and low temperature observed in the Sahara and heavy precipitations over Libya (Knipperts and Fink 2006). Figure 6.2 shows the simulated dust loading for 4th March 2004 at 12 UTC. To investigate the impact of the dust aerosols on radiation two simulations were performed; one with no interaction between the actual aerosol concentration and radiation, and another that takes into account the interaction.

Figure 6.3 shows the results for the shortwave radiation balance at the surface for both cases. The high dust load of the atmosphere leads to a strong modification of the shortwave radiation balance. In contrast to the simulation without interaction between aerosols and radiation the simulation using the actual values of the dust

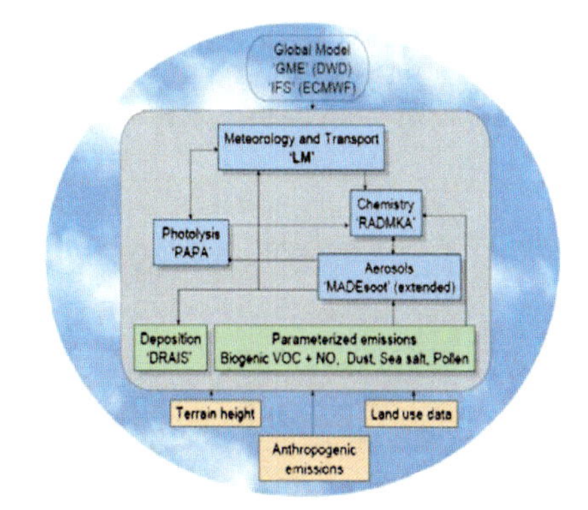

Fig. 6.1 The model system COSMO-ART

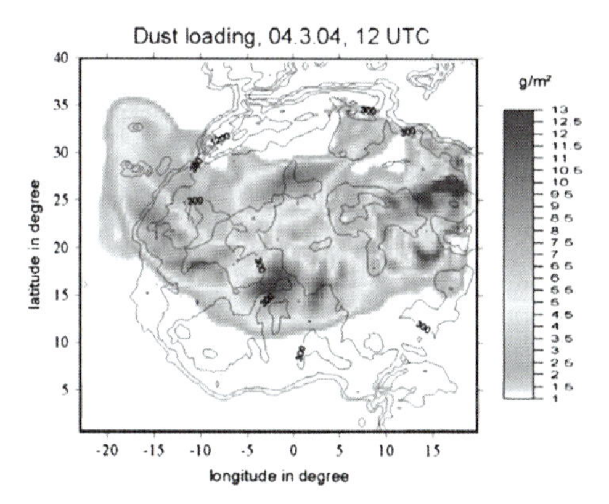

Fig. 6.2 Horizontal distribution of the simulated dust loading over western Africa

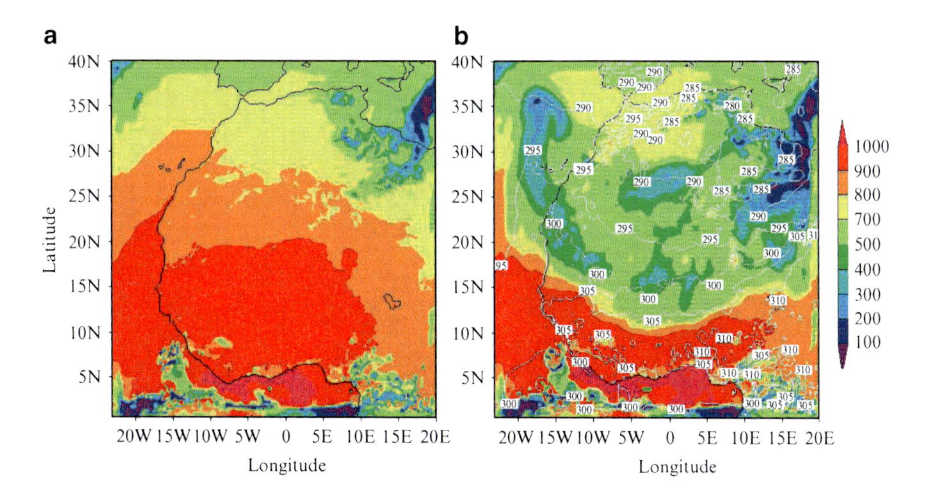

Fig. 6.3 Horizontal distribution of the according shortwave radiation balance for the simulation taken into account the actual dust concentration (**a**) and the simulation using the climatological values (**b**), both for 04.03.2004, 12 UTC

concentrations in the radiation scheme shows a reduced shortwave radiation balance up to a factor of 2 over western Africa, which has also an effect on the cloud formation and the dynamics (not shown here).

6.3.2 *The Interaction of Anthropogenic Aerosols with Radiation*

The model system was also applied to study the direct effect of anthropogenic aerosols on radiation. In addition the ageing process of the emitted soot particles was taken into account. The model domain for this study covered the south western part of Germany with adjacent areas. The simulation period was from 16.08.2005 to 22.08.2005. The emission data were available for this period with a temporal resolution of 1 h and a horizontal resolution of 7 km. As in the previous section, two model runs were carried out; one with the interaction of the actual aerosol concentration and the radiation, and one without. Figures 6.4–6.5 show results of these simulations. In contrast to the case study of the dust event the aerosol concentration is rather low. Nevertheless, the influence on the shortwave radiation balance is quite large due to the effect that the cloud formation is also influenced by the modifications in the radiation field caused by the aerosols. Although the aerosol concentration is much lower in the western part of the domain than in the eastern part (Fig. 6.4a), an effect on the shortwave radiation balance can be seen all over the model domain (Fig. 6.5a). In the less cloudy western part, there is a cooling effect of several tenth degrees dominating, whereas in the western part both areas with warming and cooling effects can be seen (Fig. 6.5b)

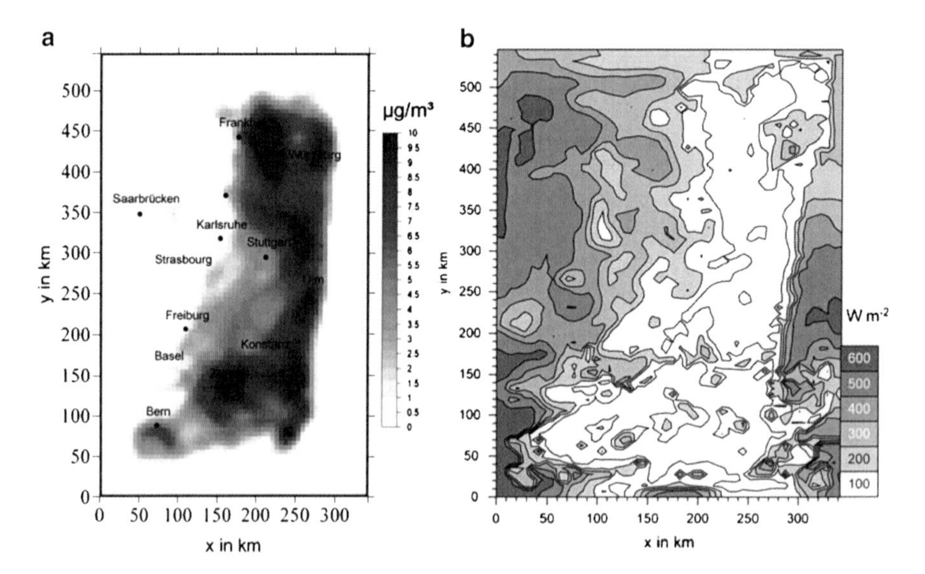

Fig. 6.4 Horizontal distribution of the dry aerosol mass (**a**) (at 20 m above ground) and the according shortwave radiation balance (**b**), both for 20.08.2005, 12 UTC

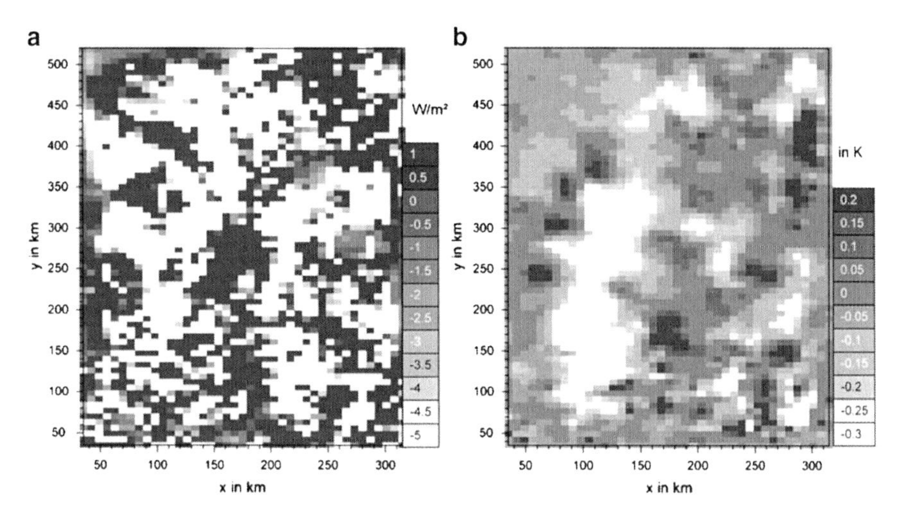

Fig. 6.5 Difference of the shortwave radiation balance (**a**) and the temperature (**b**) between the model run with aerosol–radiation interaction and the run without aerosol–radiation interaction

6.4 Conclusions

A new on-line coupled model system named COSMO LM-ART was developed. This model system contains, for example, a variety of natural and anthropogenic aerosols. The ageing process of soot is explicitly described. The treatment of their impact on the atmospheric radiation allows the quantification of feedback mechanisms. The simulation of a dust event occurring over West Africa gives rather high aerosol concentrations and consequently a strong effect on the shortwave radiation balance which leads also to differences in the cloud formation and the dynamics. The study concerning the anthropogenic aerosol–radiation-interaction shows that despite of rather low aerosol concentrations the modification of the radiation balance causes a surprisingly strong effect on the cloud pattern that needs further investigations.

Acknowledgements The authors thank their colleagues of the German Weather service for continuing support.

References

Bäumer D, Vogel B, Kottmeier Ch (2004) Parameterizing optical properties of soot-containing aerosols. European Aerosol Conf., Budapest, H, September 6–10, 2004, J. Aerosol Science, 35, pp 1195–1196

IPCC (2007) Climate change 2007: the physical science basis, Cambridge University Press, UK (Fourth Assessment Report)

Riemer N, Vogel H, Vogel B, Fiedler F (2003) Modelling aerosols on the mesoscale-μ: treatment of soot aerosol and its radiative effects. J Geophys Res 109:4601. doi:10.1029/2003JD003448
Ritter B, Geleyn J-F (1992) A comprehensive radiation scheme for numerical weather prediction models with potential applications in climate simulations. Mon Weather Rev 120:303–325
Steppeler J, Doms G, Schättler U, Bitzer HW, Gassmann A, Damrath U, Gregoric G (2003) Meso-gamma scale forecasts using the nonhydrostatic model LM. Meteorol Atmos Phys 82:75–96
Vogel B, Hoose C, Vogel H, Kottmeier Ch (2006a) A model of dust transport applied to the Dead Sea area. Meteorol Z 14:611–624
Vogel H, Vogel B, Kottmeier Ch (2006b) Modelling of pollen dispersion with a weather forecast model system. Proceedings of 28th NATO/CCMS Int. Meeting on Air Pollution Modelling and its Application, Leipzig

Chapter 7
The On-Line Coupled Mesoscale Climate–Chemistry Model MCCM: A Modelling Tool for Short Episodes as well as for Climate Periods

Peter Suppan, R. Forkel, and E. Haas

7.1 Introduction

Although on-line coupled models where meteorological and atmospheric chemistry processes are computed within one single model exist already since the 1990s and even earlier, off-line air quality models (where the chemical processes are treated independently of the meteorological model) are still widely used because of lower computational costs. However, due to this separation of meteorology and chemistry there can be a loss of possibly important information of atmospheric processes, as the meteorological information is transferred to the Atmospheric Chemistry Transport Model (ACTM) e.g. once or twice per hour. The simulation of atmospheric chemistry with an on-line coupled model can be regarded as more consistent than an off-line treatment, as the chemistry part of the model receives all necessary meteorological information directly from the meteorological part of the model at each time step without any temporal interpolation. Especially on the regional scale with grid sizes down to 1 km, the wind field and other meteorological parameters are highly variable and neglecting these variances may introduce certain errors. Although the advantages of on-line coupled meteorology–chemistry simulations against an off-line treatment are most effective for fine horizontal resolutions, effects already become significant at horizontal resolutions of around 30 km (Grell et al. 2004).

7.2 Description of MCCM

The on-line coupled regional meteorology–chemistry model MCCM (Mesoscale climate chemistry model, Grell et al. 2000) has been developed at the IMK-IFU on the basis of the non hydrostatic NCAR/Penn State University mesoscale model

P. Suppan (✉)
Institute for Meteorology and Climate Research (IMK-IFU), Karlsruhe Institute of Technology (KIT), Garmisch-Partenkirchen, Germany
e-mail: peter.suppan@kit.edu

A. Baklanov et al. (eds.), *Integrated Systems of Meso-Meteorological and Chemical Transport Models*, DOI 10.1007/978-3-642-13980-2_7,
© Springer-Verlag Berlin Heidelberg 2011

MM5 (Grell et al. 1994). The full coupling of meteorology and chemistry ensures that the air quality component of MCCM is fully consistent with the meteorological component. Both components use the same transport scheme, grid, and physics schemes for subgrid-scale transport. Similar to MM5 the MCCM model can be applied over a range of spatial scale from the regional (several thousand kilometers, resolution of 30–100 km) to the urban (100–200 km, resolution of 1–5 km) scales.

MCCM includes several tropospheric gas phase chemistry modules (RADM, RACM, RACM-MIM (Stockwell et al. 1990, 1997; Geiger et al. 2003)) and a photolysis module. Optional aerosol processes are described with the modal MADE/SORGAM aerosol module (Schell et al. 2001) which considers as single compounds sulphate, nitrate, ammonium, water, and four organic compounds. For the Aitken and the accumulation modes the gas/particle phase partitioning of the secondary sulphate/nitrate/ammonium/water aerosol compounds is based on equilibrium thermodynamics. The organic chemistry assumes that secondary organic aerosol compounds (SOA) interact with the gas phase and form a quasi-ideal solution.

Biogenic VOC and NO emissions are calculated on-line based on land use data, simulated surface temperature and radiation. Anthropogenic emissions of primary pollutants, like NO_x, SO_2, and hydrocarbons, as well as emissions of primary particulate matter have to be supplied either at hourly intervals or as yearly data from gridded emission inventories. Validation studies with MCCM have shown its ability to reproduce observed meteorological quantities and pollutant concentrations for different conditions and regions of the Earth (Forkel and Knoche 2006; Forkel et al. 2004; Grell et al. 1998, 2000; Jazcilevich et al. 2003; Kim and Stockwell 2007; Suppan and Skouloudis 2003; Suppan and Schädler 2004; Suppan 2010).

Furthermore, the model is linked to other models like hydrological or/and biosphere based models in order to describe the interactions and feedback mechanisms from each compartment to the air quality and vice versa. A schematic description of the model is given in Fig. 7.1.

The following examples show different applications of MCCM for short period simulations on air quality and emission reductions scenarios as well as simulations performed for the assessment of the impact from climate change to the regional air quality.

7.3 Applications

7.3.1 Evaluation Studies

In order to evaluate the performance of the three chemistry mechanisms included in MCCM, the results of simulations over a 2 month period in summer 2003 were compared with observations.

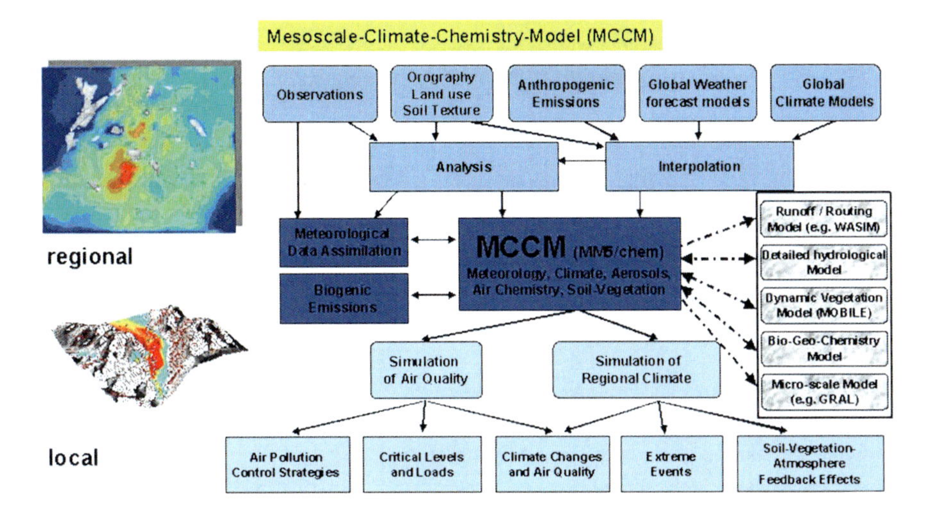

Fig. 7.1 Schematic description of MCCM as used on the local and regional scale for air quality simulations, and for simulation of the regional climate including the introduced coupled models to hydrology and biosphere compartments

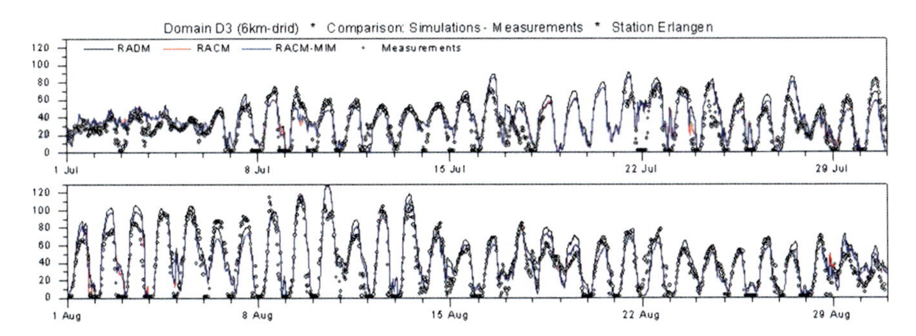

Fig. 7.2 Calculated and measured time series of ozone at the Erlangen station for three different chemical schemes

The mechanisms were constructed using a software engineering tool for chemistry kinetics (Damian et al. 2002). By using a pre-processor for the construction of the numerical integration code, the solver (numerical integration scheme) and the underlying mechanisms were decoupled and by using identical integrations schemes, the influence of the numerics to the species evolution is comparable. The simulations were performed with four nested domains (54, 18, 6 and 2 km grid resolution). They cover the time period from 1st July to 31st August 2003, and include episodes with high photosmog concentrations. The simulated species mixing ratios were compared with observations from the Bavarian measurement network (LFU). The evolution of the time series of the measured (LFU-Station Erlangen) and the simulated ozone (at 6 km resolution) are shown in Fig. 7.2. As seen, this figure shows a good correlation between measurements and simulations. But on the other side, it is also seen that the RADM mechanism overpredicted ozone concentrations during the day, while the RACM and RACM-MIM simulation results fit better to the observations (Haas et al. 2010).

7.3.2 Air Quality Studies

7.3.2.1 Effect of Highway Emissions

A typical application for assessing the influence of specific emission sources to the air quality is demonstrated within the next example. To assess the influence of highway emissions to the ozone and nitrogen dioxide concentration fields, the line source (highway, ca. 35 km) between Munich and Augsburg was excluded and the emissions were set to the surrounding levels. This could be attributed to a reduction of 80% of the traffic emissions.

As seen in Fig. 7.3, the increase of O_3 concentrations was up to 10%, whereas the NO_2 concentrations showed a decrease of 25%. The red line marks the region of influence of this specific emission reduction.

Compared to the NO_2-influenced region, O_3 has a smaller impact on the region (less than 800 km^2) but the impact is lasting longer (more than 12 h). The NO_2 influenced regions (close to 1,000 km^2) are larger, but they existed for a shorter period (less than 6 h). The main time interval of influence for both NO and O_3 occurs during the night (Suppan and Schädler 2004).

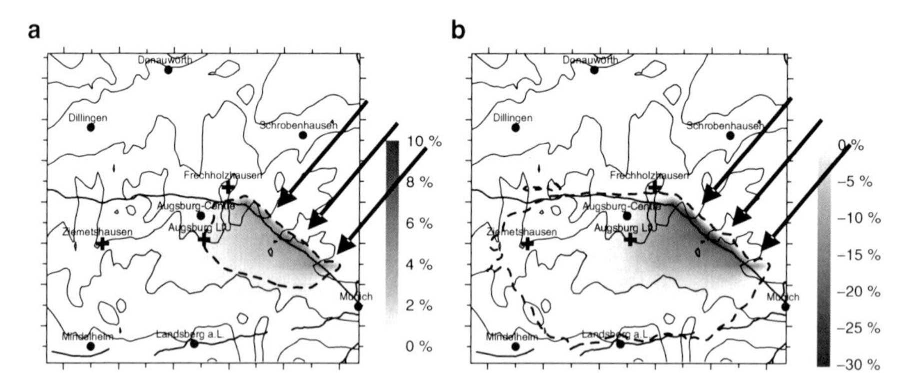

Fig. 7.3 Traffic emission effect to O_3 (**a**) and NO_2 (**b**) concentrations at the nearby region of a highway during a 4 day period with strong north-easterly winds. The thick *dotted black line* indicates the region of influence

7.3.2.2 Scenario Simulations for Mexico City

Mexico City suffers from severe air quality problems with maximum ozone values up to 250 ppb. In order to study the effect of different precursor emissions and possible mitigation measures on the ozone levels in Mexico City the MCCM was applied in the Greater Area of the city (Jazcilevich et al. 2003; Forkel et al. 2004). Figure 7.4a shows that, in agreement with observations, the simulated O_3 maxima occur in the southwest of the city (measurement station PED), which is down-wind of the city centre, as an uphill flow is prevailing during the afternoon. The minimum ozone concentrations are found in the centre of the city (station MER), where the NO emitted by traffic titrates the ozone, and in the northern part of the city at places where the NO emissions from industry and power plants locally reduce the ozone concentrations (near station XAL).

Compared to the ozone concentrations predicted for the 2010 baseline emissions, the emissions for a mitigation scenario including the replacement of old private cars, low sulphur diesel standards, the replacement of microbuses, and the relocation of two power plants (corresponding to a reduction of anthropogenic emissions of NO_x by about 20% and of VOC by 10%) result in a decrease of the daytime ozone concentrations between 5 and 25 ppb at most locations. However, in the centre of the city and for the locations where power plants are switched off, the noontime ozone concentrations are higher for the mitigation scenario than for the baseline case since less ozone is titrated in case of the reduced NO emissions at these locations. The model results indicate that taken the projected emissions for 2010, extremely strong emission reduction measures for Mexico City would be necessary in order to significantly improve the air quality in the city.

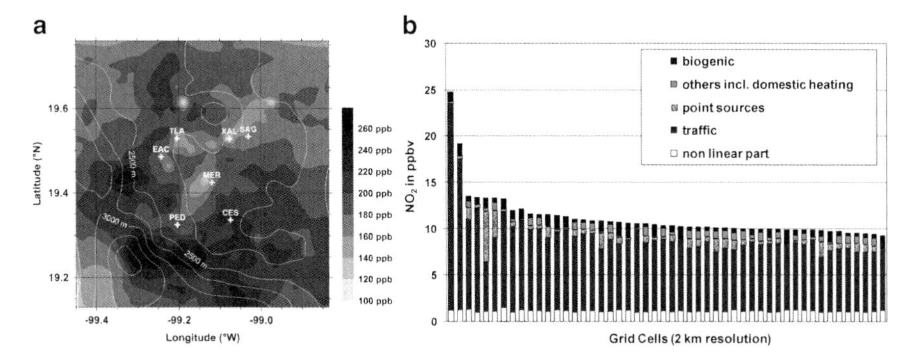

Fig. 7.4 (**a**) Average surface ozone concentration (in ppb) over 8 days at 16:00. Mexican summer time for the baseline 2010 emission scenario. (**b**) Source-receptor relationship for the 50 grid cells with highest NO_2-concentrations within the conurbation of Munich during a 4 day period

7.3.2.3 Source Receptor Analysis

Source–receptor analysis was performed to allocate the air chemistry parameters to the individual emission sources. To estimate the impact of a source group on a certain pollutant, several simulations have to be accomplished. To minimize the associated uncertainties (non-linearity of chemical processes), the source group was suppressed. Due to the non-linear chemical processes, background concentrations and advection a "non-linear" fraction has to be introduced (DG-ENV 2001). The source-receptor analysis is an important tool for abatement and emission reduction strategies.

In Fig. 7.4b the source–receptor distribution for NO_2 is shown for a short time period in the greater area of Munich/Germany. The concentration at each grid cell includes several source categories. In accordance with the NO_x emission distribution the NO_2 concentrations caused by the traffic show also the highest values (Suppan 2010).

7.3.2.4 Climate Chemistry Simulations

In order to investigate possible effects of global climate change on the near-surface concentrations of photochemical compounds in southern Germany, nested regional simulations with MCCM were carried out (Forkel and Knoche 2006). The simulations with horizontal resolutions of 60 (for whole Europe) and 20 km (for central

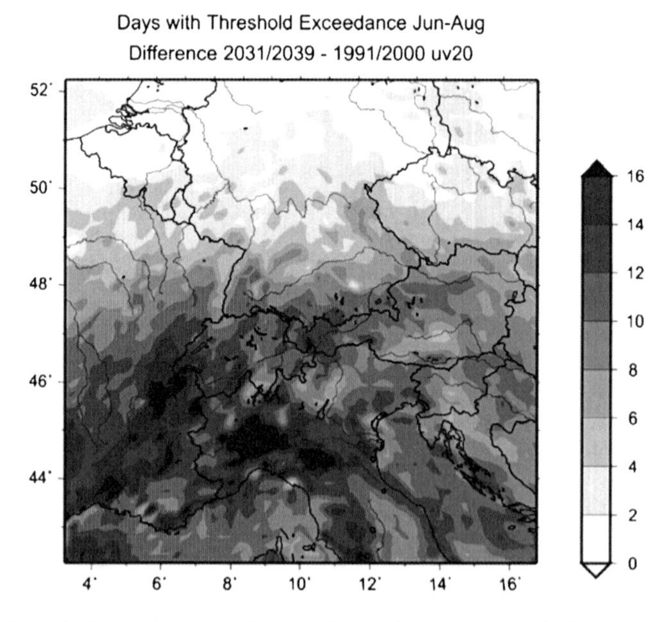

Fig. 7.5 Effect of climate change on the exceedance of the ozone threshold value

Europe) were driven by meteorological boundary conditions provided by a long-term simulation of the global climate model ECHAM4. Two time slices (representing 1990s and 2030s) of about 10 years each were compared.

For the region of southern Germany the simulations show an increase of the mean summer temperature by almost 2°C along with a decrease of cloud water and ice and a corresponding increase of the photolysis frequencies and the emissions of biogenic hydrocarbons. Under the model assumption of unchanged anthropogenic emissions this leads to an increase of the mean mixing ratios of most photooxidants. Because of the complex topography and the heterogeneous distribution of precursor emissions all parameters show pronounced regional patterns. The average daily maximum ozone concentrations in southern Germany increase for the selected scenario by nearly 10% in summer months. Depending on the region the increase of the mean daily maximum ranges between 2 and 6 ppb. As a consequence, the number of days when the 8-h mean of the ozone concentration exceeds the threshold value of 120 $\mu g/m^3$ increases by 5–12 days per year (Fig. 7.5).

7.4 Conclusions

The on-line coupled model MCCM (which is based on well known and validated model MM5) has demonstrated its applicability to support and to address air quality issues, like emission reduction scenarios, abatement strategies, or the impact of climate change to the air quality in sensitive and urbanized areas both on the regional and local scales. Furthermore, this model is able to couple or to link also other models from different compartments like hydrology or biosphere.

References

Damian V, Sandu A, Damian M, Potra F, Carmichael GR (2002) The Kinetic PreProcessor KPP – a software environment for solving chemical kinetics. Comput Chem Eng 26(11):1567–1579

DG-ENV (2001) The AUTOOIL-II programme: Air-Quality report, EC DG-ENV, edited at JRC Ispra, Report EUR 19725 EN, version 7.2 and http://europa.eu.int/comm/environment/autooil/index.htm

Forkel R, Smiatek G, Hernandez F, Iniestra R, Rappenglück B, Steinbrecher R (2004) Numerical simulations of ozone level scenarios for Mexico City. 84th AMS Annual Meeting, 6th conference on atmospheric chemistry: air quality in megacities. Seattle, WA. 11–15 January 2004, Combined Preprint CD, contribution P1.2 (4p.), http://ams.confex.com/ams/84Annual/techprogram/paper 70640.htm

Forkel R, Knoche R (2006) Regional climate change and its impact on photooxidant concentrations in southern Germany: simulations with a coupled regional climate-chemistry model. J Geophys Res 111(D12):13pp, G12302, doi:10.1029/2005JD006748

Geiger H, Barnes I, Benjan I, Benter T, Splitter M (2003) The tropospheric degradation of isoprene: an updated module for the regional chemistry mechanism. Atmos Environ 37:1503–1519

Grell GA, Dudhia J, Stauffer DR (1994) A description of the fifth-generation Penn State/NCAR Mesoscale Model (MM5). NCAR Tech Note TN-398 + STR, 122 p

Grell GA, Schade L, Knoche R, Pfeiffer A (1998) Regionale Klimamodellierung, Final Report, Joint Proj. BayFORKLIM, Subproject K2, Fraunhofer-Institut für Atmos. Umweltforschung, Garmisch-Partenkirchen, Germany

Grell GA, Emeis S, Stockwell WR, Schoenemeyer T, Forkel R, Michalakes J, Knoche R, Seidl W (2000) Application of a multiscale, coupled MM5/Chemistry Model to the complex terrain of the VOTALP Valley Campaign. Atmos Environ 34:1435–1453

Grell GA, Knoche R, Peckham SE, McKeen SA (2004) Online versus offline air quality modelling on cloud-resolving scales. Geophys Res Lett 31:4

Haas E, Forkel R, Suppan P (2010) Application and Inter-comparison of the RADM2 and RACM Chemistry Mechanism including a new Isoprene Degradation Scheme within the Regional Meteorology-Chemistry-Model MCCM. Int J Environ Pollut 40(1–3):136–148

Jazcilevich AD, Garcia AR, Ruiz-Suarez LG (2003) A study of air flow patterns affecting pollutant concentrations in the Central Region of Mexico. Atmos Environ 37:183–193

Kim D, Stockwell WR (2007) An online coupled meteorological and air quality modelling study of the effect of complex terrain on the regional transport and transformation of air pollutants over the Western United States. Atmos Environ 41:2319–2334, 16 pp

Schell B, Ackermann IJ, Hass H, Binkowski FS, Ebel A (2001) Modelling the formation of secondary organic aerosol within a comprehensive air quality model system. J Geophys Res 106:28275–28293

Stockwell W, Middelton P, Chang J (1990) The second generation regional acid deposition model – chemical Mechanism for regional air quality modelling. J Geophys Res 95:16343–18367

Stockwell W, Kirchner F, Kuhn M, Seefeld S (1997) A new mechanism for region atmospheric chemistry modelling. J Geophys Res 102:847–879

Suppan P, Skouloudis A (2003) Inter-comparison of two air quality modelling systems for a case study in Berlin. Int J Environ Pollut 20:75–84

Suppan P, Schädler G (2004) The impact of highway emissions on ozone and nitrogen oxide levels during specific meteorological conditions. Sci Total Environ 334:215–222

Suppan P (2010) Assessment of air pollution in the conurbation of Munich – present and future. Int J Environment and Pollution 40:1,2,3(149–159)

Chapter 8
BOLCHEM: An Integrated System for Atmospheric Dynamics and Composition

Alberto Maurizi, Massimo D'Isidoro, and Mihaela Mircea

8.1 Introduction

Chemical composition at regional scale is a subject of increasing interest for both air quality and climatological issues. Several models exist, but no on-line Italian model was available some years ago. Different expertises are presented at ISAC-CNR on several aspects involved in air quality modelling: atmospheric chemistry, meteorology, microphysics and turbulence, and these were gathered to develop BOLCHEM. This model is a part of the European community project GEMS, sub-project Regional Air Quality models (RAQ), and it has been used to study different aspects of air quality related problems.

8.2 Model Description

The BOLCHEM model (BOLam + CHEMistry) is the result of an on-line coupling between the mesoscale meteorological model BOLAM (BOlogna Limited Area Model http://www.isac.cnr.it/~dinamica/bolam/index.html) (Buzzi et al. 1994, 2003) and modules for transport and transformation of chemical species. BOLAM dynamics is based on hydrostatic primitive equations, with wind components, potential temperature, specific humidity, surface pressure, as dependent variables. The vertical coordinate system is hybrid-terrain-following, with variables distributed on a non-uniformly spaced staggered Lorenz grid. The horizontal discretisation uses geographical coordinates on the Arakawa C-grid. The time scheme is split-explicit, forward-backward for gravity modes. A 3D WAF (Weighted Average Flux) advection scheme coupled with semi-Lagrangian advection of hydrometeors is implemented. A fourth order horizontal diffusion of the prognostic variables (except for Ps), a second divergence diffusion and damping

A. Maurizi (✉)
Institute of Atmospheric Sciences and Climate, Italian National Research Council, Bologna, Italy
e-mail: a.maurizi@isac.cnr.it

A. Baklanov et al. (eds.), *Integrated Systems of Meso-Meteorological and Chemical Transport Models*, DOI 10.1007/978-3-642-13980-2_8,
© Springer-Verlag Berlin Heidelberg 2011

of the external gravity mode are included. The lateral boundary conditions are imposed using a relaxation scheme that minimizes wave energy reflection. The initial and lateral boundary conditions are supplied from the ECMWF (European Centre for Medium-range Weather Forecasts) analyses available at $0.5^\circ \times 0.5^\circ$ horizontal resolution. Hybrid model level data are directly interpolated on the BOLAM grid.

Transport (advection and diffusion) of tracers (both passive and reactive) is performed on-line at each meteorological time-step using WAF scheme for advection and a "true" (second order) diffusion, with diffusion coefficient carefully estimated from experiments (Tampieri and Maurizi 2007). Vertical diffusion is performed using 1D diffusion equation with a diffusion coefficient estimated by means of an *k-l* turbulence closure scheme. Dry deposition is computed through the resistance-analogy scheme and is provided as a boundary condition to the vertical diffusion equation. Furthermore, vertical redistribution of tracers due to moist convection is parameterized consistently with the Kain-Frisch scheme used in the meteorological part for moist convection. Transport of chemical species is performed in mass units while gas chemistry is computed in ppm.

Physical/chemical processes are treated separately for the gas phase, aerosol classes, and generic tracers (e.g. radioactive species, Saharan dust, etc.). Gas phase is treated using the SAPRC90 or CB4 chemical mechanisms. Aerosols are modeled using M7 module from ECHAM5 (coupling is still in progress) and generic species are defined by the user case by case providing chemical/physical properties and equations. More technical details can be found in the COST-728/732 model inventory: http://www.mi.uni-hamburg.de/List_classification_and_detail_view_of_model_entr.567.0.html?&user_cost728_pi2 [showUid]=80.

8.3 Model Applications

The model has been used for a variety of situations in order to test the reliability of the choices made. It also currently runs at ECMWF in the frame of the GEMS Project for the ensemble near-real-time experiment. Some of the main results are briefly reported in the following sections.

8.3.1 Evaluation of Model Performances for Ozone

The performances of BOLCHEM on the ability to predict O_3 concentration over Italy were evaluated (Mircea et al., 2008). The comparison between computed and measured concentrations for some periods of 1999 showed that the model is capable to predict the diurnal cycle of O_3, in particular in summer. The agreement between modeled and measured quantities is good during daytime while at night there is some problem connected to O_3 destruction. However, US-EPA's criteria are met;

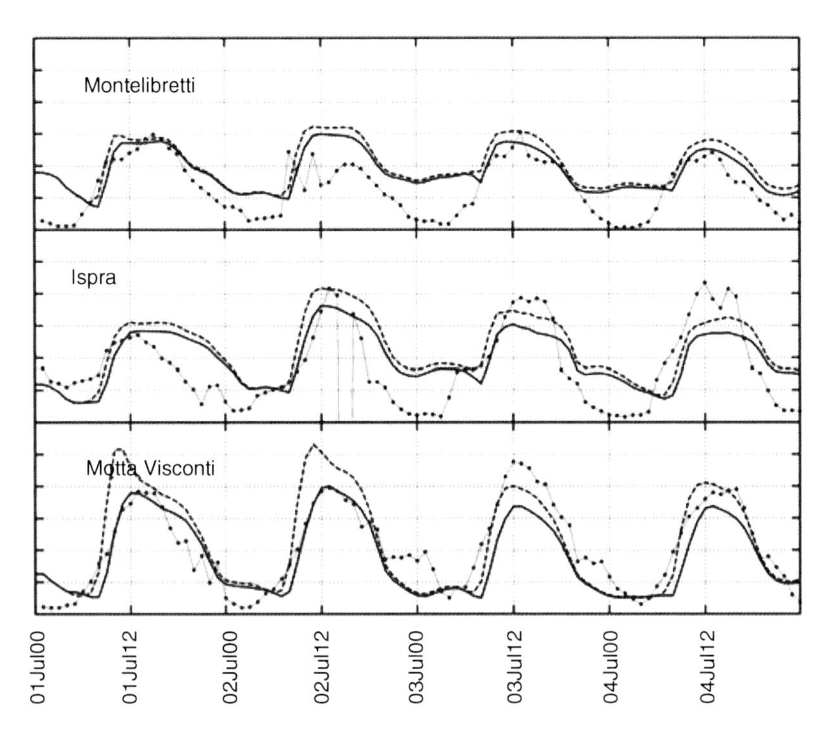

Fig. 8.1 Time series for ozone at three Italian locations: Ispra, Motta Visconti (MI) and Montelibretti (RM) for 4 months: January, June, July and August

so, that model results can be reliably used for air quality predictions. Some time series of O_3 computed with CBIV and SAPRC90 mechanisms compared with measurements are shown in Fig. 8.1.

8.3.2 Ozone Sensitivity to Precursor Emission Reduction

An important aspect for emission reduction policies is the study of the regional sensitivity to precursor emission reduction. For this purpose indicator species are computed to asses reduction in NOx and VOC over the whole Italy and on subdomains centered on specific spots: the Milan and Rome areas and some of the major industrial areas. Different periods were selected, and runs with both chemical schemes were performed over Italy comparing indicator species computed for different reduction scenarios. For all the periods investigated, it was found that Italy, including the large islands Sicily and Sardinia, is mostly dominated by NOx chemical regimes, independently of the photochemical mechanism used. However, the effect of the NOx reduction predicted with the CB-IV mechanism is lower than that predicted with SAPRC90 mechanism. In addition, the urban areas

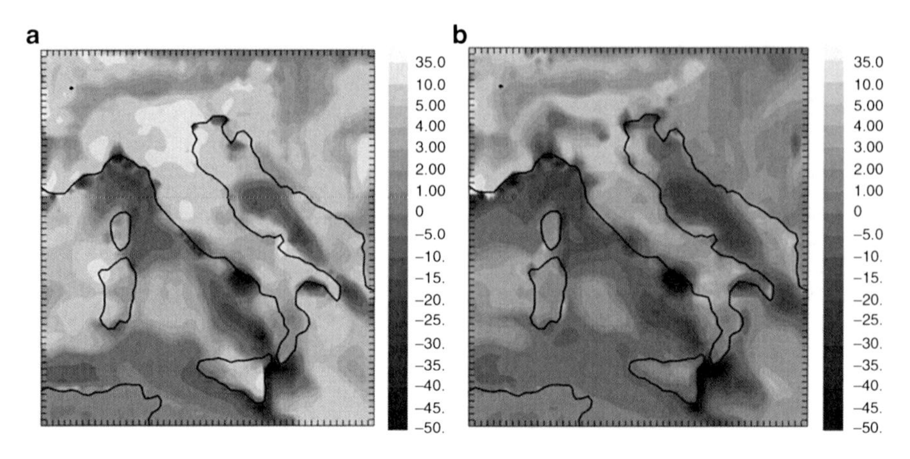

Fig. 8.2 Maps of sensitivity of ozone to reduction of VOC (positive) and NOx (negative). Left panel SAPRC90, right panel CBIV

of cities Milan, Rome, Naples or/and industrial areas around harbors of Genoa, Messina, and Venice are always in a marked VOC sensitive regime. Note that differences in the spatial distribution of the chemical regimes due to the photochemical mechanism used and due to the meteorological conditions are comparable. Examples of ozone sensitivity maps to reduction of VOC and NOx are presented in Fig. 8.2.

8.3.3 Saharan Dust Transport

Saharan dust is a major component of the aerosol load over Italy due to vicinity to the African continent. The direct forcing of dust aerosol may be comparable to or even exceed the forcing of anthropogenic aerosols. To correctly treat this aspect, a proper modelling of dust blowing sources is needed. A sensitivity experiment was carried out to test the sensitivity of the emission model (Tegen et al. 2002) to the friction velocity threshold during a strong Saharan dust outbreak that occurred from 15 to 19 July 2003 (Mircea et al., 2007), transporting the dust particles almost over the whole Italy (Fig. 8.3). The comparison of model results with the observations (surface concentrations from EMEP stations and aerosol optical depth (AOD) from AERONET stations) allowed selecting the better threshold.

8.3.4 Lagrangian Transport and Etna Eruption

A Lagrangian transport model implemented in BOLCHEM (BOLTRAJ variant) can be used in conjunction with the Eulerian part (D'Isidoro et al., 2005). This is

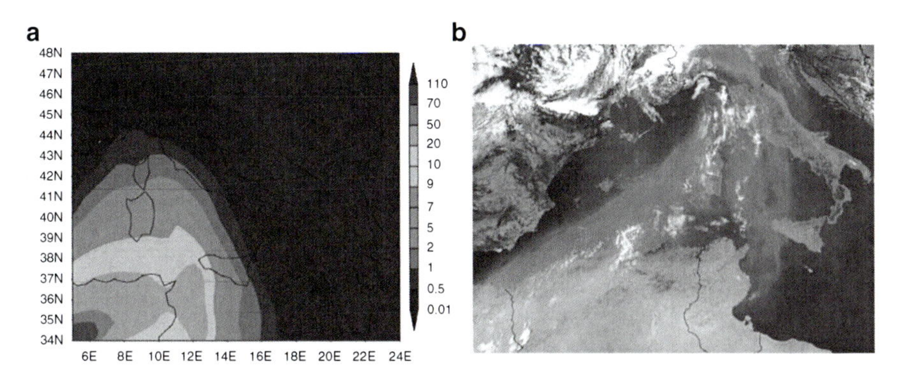

Fig. 8.3 Vertically integrated aerosol load simulated by (**a**) BOLCHEM and (**b**) seen by MODIS AQUA

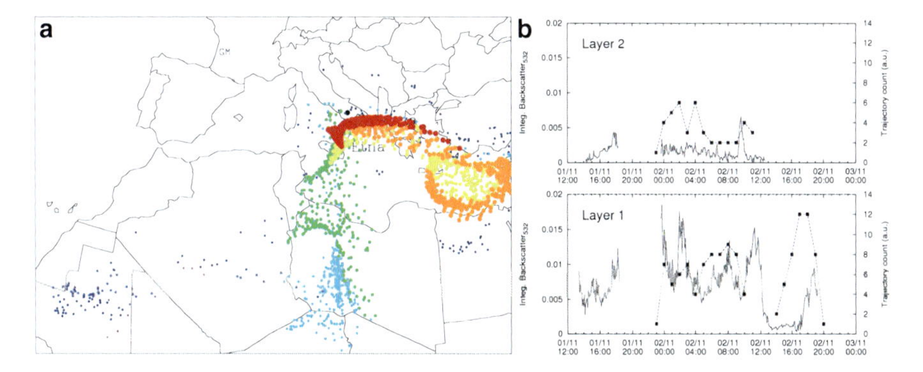

Fig. 8.4 (**a**) Position of particles released at Mt. Etna from 26-10-2002 to 01-11-2002. Different colours scales denote different ages (in days); (**b**) Comparison of LIDAR signal measured over Potenza and simulated by BOLCHEM. Continuous line: measured; dotted line: modelled

useful as an analysis tool for a better interpretation of Eulerian simulations through the computation of the probability matrices of pollutant origin. It is also useful to study specific events of concentrated point sources when the resolution of the Eulerian grid would be too small to represent dispersion at short time (Tiesi et al., 2006). The first application was the Mt. Etna eruption on autumn 2002 (Villani et al. 2006). The joint analysis of Lagrangian trajectories, satellite data, meteorology, and lidar measurements allowed to estimate the tropospheric dispersion coefficient and to clarify some features observed by lidar located near Potenza. Lidar measurements recorded a strong signal with a clear sulphate signature, along with weaker layers above. Satellite images gave no clear evidence, but trajectory analysis revealed the nature of the complex picture. Part of the trajectories traveled over Potenza taking a long passage through the Sahara region, possibly carrying some silicate and passing over Potenza at the same time (and at different heights) of those coming directly from Mt. Etna (Fig. 8.4).

Acknowledgements The authors would like to thank Francesco Tampieri and Sandro Fuzzi for their support. BOLCHEM is developed thanks to EC GEMS and ACCENT Projects, the Italian MIUR project "AEROCLOUDS", and it was also supported by the Italian Ministry of Environment through the Program Italy–USA Cooperation on Science and Technology of Climate Change.

References

Buzzi A, Fantini M, Malguzzi P, Nerozzi F (1994) Validation of a limited area model in cases of Mediterranean cyclogenesis: surface fields and precipitation scores. Meteorol Atmos Phys 53:137–153

Buzzi A, D'Isidoro M, Davolio S (2003) A case study of an orographic cyclone south of the Alps during the MAP SOP. Q J R Meteor Soc 129:1795–1818

D'Isidoro M, Maurizi A, Tampieri F, Tiesi A, Villani MG (2005) Assessment of the numerical diffusion effect in the advection of passive tracer in BOLCHEM. Nuovo Cimento C 28:151–158

Mircea M, D'Isidoro M, Maurizi A, Tampieri F, Facchini MC, Decesari S, Fuzzi S (2007) Regional modeling of aerosols using the air quality model BOLCHEM: Saharan dust intrusions over Italy. EGU General Assembly, Vienna

Mircea M, D'Isidoro M, Maurizi A, Vitali L, Monforti F, Zanini G, Tampieri F (2008) A comprehensive performance evaluation of the air quality model BOLCHEM to reproduce the ozone concentrations over Italy. Atmos Environ 42(6):1169–1185

Tampieri F, Maurizi A (2007) Evaluation of the dispersion coefficient for numerical simulations of tropospheric transport. Nuovo Cimento C 30:395–406

Tegen I, Harrison SP, Kohfeld K, Colin Prentice I, Coe M, Heinmann M (2002) J Geophys Res 107:D21. doi:10.1029/2001JD000963

Tiesi A, Villani MG, D'Isidoro M, Prata AJ, Maurizi A, Tampieri F (2006) Estimation of dispersion coefficient in the troposphere from satellite images of volcanic plumes. Atmos Environ 40:628–638. doi:10.1016/j.atmosenv.2005.09.079

Villani MG, Mona L, Maurizi A, Pappalardo G, Tiesi A, Pandolfi M, D'Isidoro M, Cuomo V, Tampieri F (2006) Transport of volcanic aerosol in the troposphere: the case study of the 2002 Etna plume. JGeophys Res 111:D21102. doi:10.1029/2006JD00712

Part II
Off-Line Modelling and Interfaces

Chapter 9
Off-Line Model Integration: EU Practices, Interfaces, Possible Strategies for Harmonisation

Sandro Finardi, Alessio D'Allura, and Barbara Fay

9.1 Introduction

A survey of the multiple and diverse modelling communities in European countries was performed in COST728 on a basis of partner contributions (Baklanov et al. 2007). Even if the model coverage remains incomplete and somewhat arbitrary the contributions represent a wide spectrum of modelling complexity and efforts in 16 European countries and about 40 institutions. The majority of the presented systems are based on mesoscale meteorological models (MetMs) available at the national weather services or weather forecasting consortia (i.e. HIRLAM, COSMO (formerly Lokalmodell), ALADIN) and on international free community models developed by universities (i.e. MM5, WRF, MC2, RAMS). This approach allows the air quality (AQ) modelling community to take advantage and benefit from development, testing and model validations done for the purpose of numerical weather prediction (NWP). Moreover, it allows users without large development resources to share their experience and obtain support by a wide user and developer community.

The modelling components that deal with transport and transformation of atmospheric pollutants are more diverse than the MetMs, ranging from a simple passive tracer along a trajectory (i.e. CALPUFF) to a complex treatment of reactive gases in Earth Modelling systems (i.e. MESSy). The wide spectrum of model applications ranges from diagnostic or climatologic AQ assessment, episode analysis and source apportionment to AQ forecasting at regional and urban scales and emergency preparedness for toxic and radioactive releases.

The communication between off-line coupled meteorological and AQ models is often a problem of underestimated importance. The variety of modelling systems previously introduced give rise to different approaches and methods implemented within interface modules. Tasks covered by interfaces are minimized in coupled systems relying on surface fluxes, turbulence and dispersion parameters (i.e. eddy

S. Finardi (✉)
ARIANET s.r.l, via Gilino 9, 20128, Milano, Italy
e-mail: s.finardi@aria-net.it

A. Baklanov et al. (eds.), *Integrated Systems of Meso-Meteorological and Chemical Transport Models*, DOI 10.1007/978-3-642-13980-2_9,
© Springer-Verlag Berlin Heidelberg 2011

viscosity) provided by the meteorological driver. Other systems use interface modules implementing surface and boundary layer parameterizations to estimate dispersion parameters. Atmospheric physics parameterizations, and even default or limit values assumed for some key parameters, can have relevant effects on pollutant concentration fields in critical conditions (e.g. low wind and stable conditions). Interface modules can involve the evaluation of emissions of some relevant species that can be strongly influenced by meteorology, like biogenic VOC, pollen emission, windblown dust and sea salt spray. Moreover, mesoscale and urban scale AQ modelling systems are usually nested within larger scale application results, used to initialize and drive AQ fields at domain boundaries. This operation too can have relevant influence on AQ modeled fields, and becomes especially evident over complex topography.

9.2 Off-Line Coupled Models and Interfaces

The major components of an integrated meteorological and AQ modelling system are sketched in Fig. 9.1. Since the input data flow connects the meteorological modelling system and AQ model those two are generally defined as coupled models. Depending on the characteristics of this connection we can distinguish between off-line and on-line coupling. Off-line coupled MetMs and AQ models work separately, there is no feedback from Atmospheric Chemical Transport Models (ACTMs) to MetMs and meteorological input to the AQ model is usually limited to averages, either in time or space, of main variables defining the atmospheric status (fields are provided at any fixed times, e.g. 1 h). This specific approach is the traditional way by which those complex systems have been developed until now.

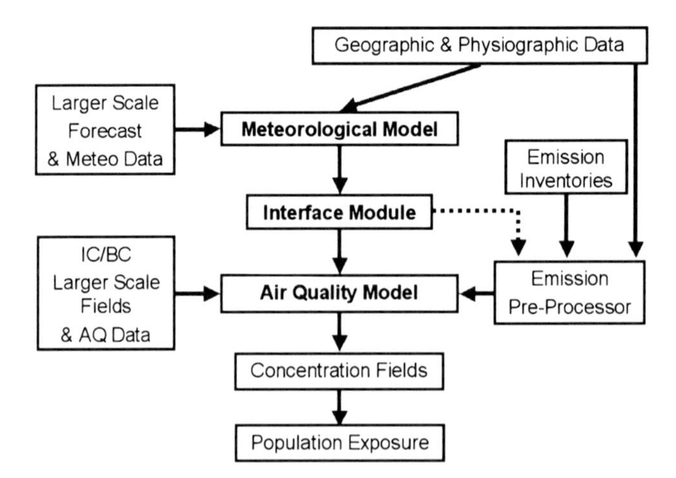

Fig. 9.1 Conceptual scheme of integrated meteorological and air quality modelling system

The development of these modelling systems is usually focused on the scientific and technical features of emission, atmospheric flow and pollutant dispersion models, while comparatively little attention is devoted to the connection of different models. Meteorological and AQ models often employ different coordinate systems and computational meshes. In principle, interfaces should simply solve this grid system mismatch to connect MetMs output and AQ models input with minimum possible data handling.

Nonetheless, interface modules are often used to solve other system realization issues, e.g.:

- Some AQ models rely on "standard" meteorological products which usually do not include turbulence, atmospheric stability, mixing height, and dispersion coefficients
- MetMs cannot provide all the physical variables that are needed by AQ models (e.g. deposition velocities) or some meteorological fields may be estimated by parametrizations not compatible with modelling methods implemented in dispersion models
- Sometimes re-computation or "filtering" of dispersion parameters is considered more robust for practical applications
- Horizontal resolution of the meteorological forecast can be lower than that needed by AQ models, and insufficient to correctly estimate dispersion parameters

To solve the above mentioned problems various tasks are often included within interface modules, as well as: data interpolation, meteorological fields downscaling, boundary layer parametrizations and estimation of dispersion coefficients, evaluation of meteorological driven emissions (e.g. biogenic, wind blown dust, sea salt), enhancement of physiographic data.

A multiplicity of off-line coupled modelling systems has been developed and applied all over the world. The most common systems features and interfacing strategies in Europe have been identified, within COST-728/WG2, using: COST-728/-732 model inventory (http://www.mi.uni-hamburg.de/index.php?id=539), a questionnaire on interfaces circulated among COST-728 participants and previous experiences and knowledge of COST-728/WG2 members (Baklanov et al. 2007).

Three main approaches have been identified:

- Joint development of coupled models, with interfaces built on specific models features and needs, this approach is mainly adopted by large institutions and Weather Services developing both MetM and AQ models
- Use or customization of US Community modelling systems, e.g. MM5/WRF +CMAQ with MCIP interface module
- Interfacing of self developed AQ models with EU Weather Services and US Community Meteorological Models through model specific or general purpose interfaces

The first strategy implies the direct use of physical parameters estimated by MetMs, like e.g. Monin–Obukhov similarity theory (MOST) parameters, and limitation of the interface module tasks to the evaluation of missing variables. This approach is particularly attractive when meteorological and AQ models share the same computational grid system and data interpolation can be avoided. Note that changing grid system and topography makes necessary to re-compute the vertical wind component, to guarantee the mass conservation, which is essential for dispersion calculations.

The other approaches are mainly based on the development of meteorological pre-(post-) processors capable to evaluate surface and boundary layer scaling parameters, mixing height, atmospheric turbulence and dispersion parameters on the basis of the average meteorological variables provided by MetMs and possibly supplementary external data. This approach gives the possibility to interface meteorological and AQ models characterized by relevant differences that can make a direct connection difficult. Moreover, it can allow the introduction of additional high resolution information, like land-use, roughness length, or urbanized parameterization to be used by computations performed in the interface module. The use of boundary layer and dispersion parameterizations within interface modules or AQ models should take into account the effective resolution of MetMs to avoid parameterization of phenomena explicitly described by modelled meteorological fields, as it can happen when high resolution MetMs results are available.

The AQ models have to be interfaced with pollutant emissions and initial and boundary conditions imposed from a larger scale AQ forecast. Pollutant emissions can be influenced by meteorological conditions through different kinds of processes of both anthropogenic and natural origin. Air temperature determines the amount of fuel consumption for house heating, the meteorological conditions often influence people behavior (e.g. car or public transport usage) determining some features of pollutant emissions. Meteorological conditions influence natural emission processes like surface erosion, wind blown dust resuspension or biogenic emissions of Volatile Organic Compounds (VOC) from vegetation or pollen.

The ACTMs results depend on the initial conditions and inflow of background concentrations into the computational domain. For meso and local scale AQ simulations, these conditions are usually defined from larger scale forecast results by an interface module that has to match grid and resolution differences and possibly the different chemical reactions schemes employed in the models considered.

The following sections provide a few examples of the possible effects of the different interfacing issues on the AQ simulation results.

9.3 Air Quality Modelling System: Results

9.3.1 Interface Module and Model Nesting Effects on Air Quality Simulation

An integrated AQ modelling system similar to those previously sketched has been applied in two different urban environments with the intent to highlight its sensitivity to dispersion process parameterizations and AQ model initialization implemented within the interface module. The modelling system, used for the following applications, is based on the MetM RAMS (Pielke et al. 1992; Cotton et al. 2003) and on the Eulerian chemical transport model FARM (Calori and Silibello 2004). The cited meteorological and AQ models are connected by the interface module GAP/SurfPRO (Calori et al. 2005; Baklanov et al. 2005). The Grid AdaPtor (GAP) is a grid interpolation tool with a capability to re-compute vertical velocities, it has been developed to interface FARM with any MetM. The SUR-Face-atmosphere interface PROcessor (SURFPRO) is a meteorological processor based on MOST designed to provide turbulence and dispersion scaling parameters, as well as eddy diffusivities and deposition velocities (Beljaars and Holtslag 1991; Hanna and Chang 1992; Zilitinkevich et al. 2002).

The cities of Rome and Turin are respectively the first and fourth largest urbanized areas in Italy. They are exposed to severe air pollution episodes induced by complex air flow patterns and atmospheric boundary layer dynamics, due to both the complexity of the urban canopy and specific mesoscale flow features (e.g. sea breezes, catabatic flows, and air stagnation). The nested computational domains for both mentioned systems are depicted in Fig. 9.2. Rome is the largest Italian city, characterized by a widely spread urbanized area, with a total population of around 3.5 million. The city is often affected by high ozone and PM concentrations, likely to be detected in both summer and winter. During summer, the high insulation favors photochemical activities. In winter, persistent high pressure systems with very weak pressure gradients determine low wind conditions and possibly temperature inversion causing pollutant accumulation in the lower layers of the troposphere.

Turin metropolitan area has a resident population of about 1.5 million. It represents the core of one of the major industrial areas in northern Italy. The city is located at the western edge of the Po Valley, and it is situated mainly on a flat

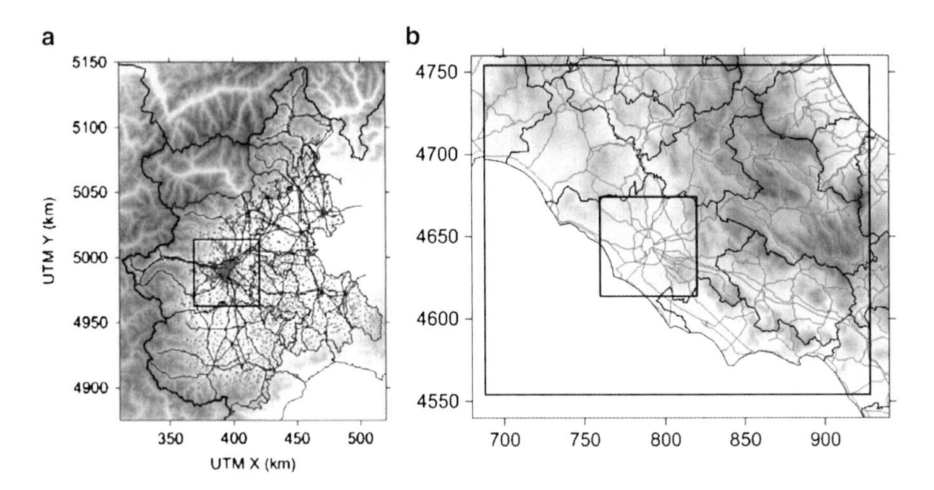

Fig. 9.2 Computational domains of Turin (**a**) and Rome (**b**) urban area air quality modelling systems

plane between the Western Alps and a range of hills on its east side. Local circulation is strongly influenced by the shelter effect of the Alpine mountains chain, and it is dominated by the superposition of mesoscale (e.g. Po Valley stagnation, mountain/valley breezes and föhn) and urban flow features.

9.3.2 Dispersion Parametrization Effect

The modelling system introduced in the previous section has been applied to analyse a summer air pollution episode in the area of Rome (Gariazzo et al. 2007). The area is exposed to sea breeze circulation during daytime, while a very week land breeze, turning to calm conditions within the city of Rome, characterizes night time circulation. Surface turbulent fluxes and Eulerian dispersion coefficients (eddy diffusivities) used by the chemical transport model FARM are computed by SURFPRO. The similarity theory is based on the general assumptions of quasi-stationary and horizontally homogeneous flow, and constant (independent of height) turbulent fluxes within the surface layer (Arya 1988). These assumptions are generally not fulfilled in urban areas and complex terrain. MOST is nevertheless applied in many models even in these cases, mostly due to the lack of other practical formulations (Mahrt 1999). During low wind conditions, like those observed in Rome at night time, the value provided by parameterizations for the eddy diffusivity is very low and usually falls below the minimum K_Z value. Different minimum values for K_Z are used by AQ models, normally ranging from 0.1 to 1 m^2s^{-1}. In SURFPRO different minimum values can be imposed as a function of land use:

$$K_z^{\min} = K_{rur}^{\min} \cdot (1 - f_{urban}) + K_{urb}^{\min} \cdot f_{urban}$$

where f_{urban} indicates the urban land use fraction within each grid cell, $K_{rur}^{\min}$ and $K_{urb}^{\min}$ indicate minima of K_Z for rural and urban areas, respectively, for which values of 0.1 and 1 m^2s^{-1} are commonly used. This assumption can be justified due to the urban canopy effect that has the tendency to maintain neutral or slightly unstable conditions over the city during the night, consequently increasing pollutant dispersion with respect to rural conditions.

Figures 9.3 and 9.4 show examples of the modelling system results for O_3 and NO_2 at an urban background (Villa Adda) and rural (Cavalieri) stations. The reference simulation (black line), with minimum K_Z defined by the previous formula, shows an overestimation of O_3 at the urban station during night-time, while NO_2 is slightly underestimated. A second simulation has been performed imposing everywhere a minimum $K_Z = 0.1$ m^2s^{-1}. The results of this run (Figs. 9.3 and 9.4, grey line) show an enhancement of O_3 at the urban station, and an excessive growth of NO_2 at night. As expected, no relevant change affects concentrations at rural location and during daytime. The stronger limitation imposed to vertical mixing by small K_Z increases NO_X concentrations and causes consumption of O_3 (ozone titration) in VOC-limited photochemical regimes.

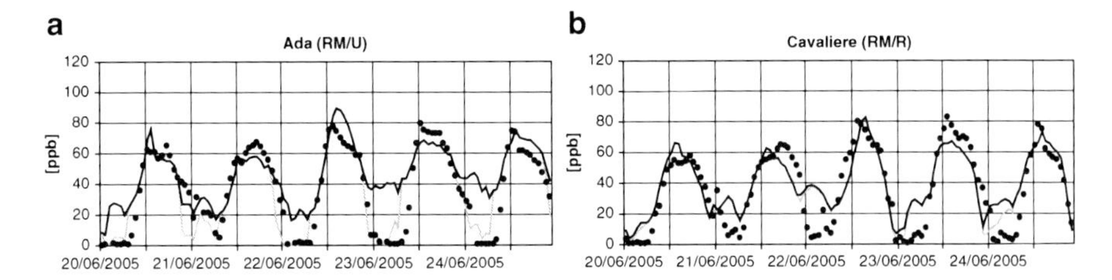

Fig. 9.3 Comparison among O_3 observed (black dots) and computed concentration with K_Z minimum value set to 0.1 m^2s^{-1} (*grey line*) and 1 m^2s^{-1} (*black line*) at urban (**a**) and rural (**b**) stations

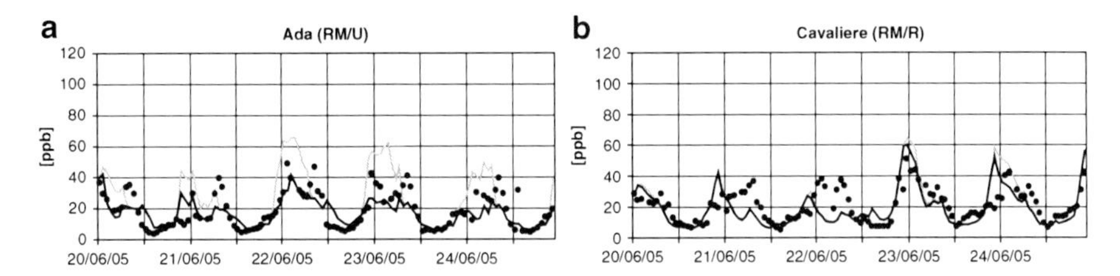

Fig. 9.4 Comparison among NO_2 observed (*black dots*) and computed concentration with K_Z minimum value set to 0.1 m^2s^{-1} (*grey line*) and 1 m^2s^{-1} (*black line*) at urban (**a**) and rural (**b**) stations

Further tests and previous experiences in other geographic locations confirmed that minimum K_Z is a relevant (and often neglected) parameter to model properly the dispersion during weak wind and very stable conditions. Unfortunately no general value for minimum K_Z can be defined, while proper values depend on season and local climatology, as well as on numerical diffusion in the advection scheme.

9.3.3 Surface Fluxes and Boundary Layer Parametrization Effect

In principle, the direct use of physical parameters estimated by MetMs should be the preferred interfacing method. The evaluation of dispersion parameters from MetMs average fields and turbulent fluxes can indeed guarantee the modelling system consistency and take advantage of new generation models capabilities, e.g. higher order turbulence closures, surface layer parameterizations and soil-surface-canopy models. On the other hand this approach can suffer the intrinsic weakness of being influenced by possible meteorological forecast errors or local scale flow features that can have relevant impact on surface fluxes, mixing height value and pollutant dispersion.

A test case to evaluate possible effects of different interfacing approaches has been run over the Torino area. The simulations covered a summer fair weather period, when thunderstorm activity occurred over the western Alps. The AQ model has been driven by two different set of turbulent surface fluxes and scaling parameter. The first set has been estimated using the surface fluxes produced by the MetM RAMS; the second – by the SURFPRO interface module employing the van Ulden and Holtslag (1985) formulation for surface fluxes and MOST.

Figure 9.5 shows the relevant differences obtained from the two test simulations for almost all the considered parameters (sensible heat flux, friction velocity, mixing height, and vertical diffusivity at the first vertical level) during the three days of simulation. The comparison of computed and observed concentrations (Fig. 9.6) highlights the mismatch of NO_2 concentrations produced by the simulation using RAMS turbulent fluxes. The sensible heat flux has been largely underestimated, producing a very limited boundary layer growth with relevant effects on local NO_X concentrations but limited influence on O_3. Further analysis pointed out that a localized convective precipitation event was mispredicted by RAMS, affecting part of Torino city with a strong precipitation event that did not occur. The high uncertainty of storm location and intensity forecast is not surprising, it is due to the geographical complexity of the region and seasonal (July) thunderstorm frequency.

During adverse meteorological events the use of an interface module to model dispersion parameters can have the advantage to reduce forecast error effects on predicted concentrations. Anyway, further analysis showed that the discussed results are strongly dependent on the radiation scheme used by RAMS model.

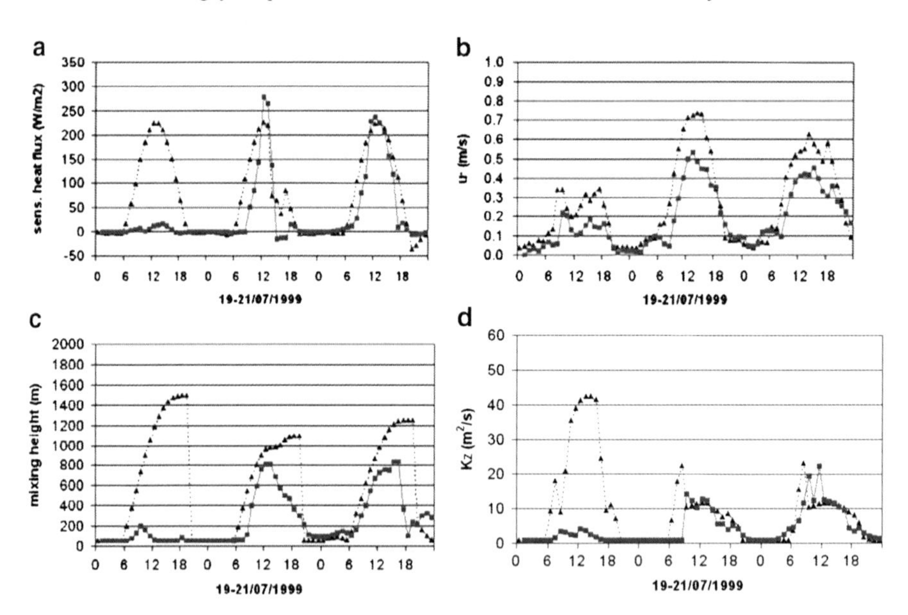

Fig. 9.5 Sensible heat flux (**a**), u* (**b**), mixing height (**c**) and K_Z (**d**) computed by RAMS (*grey solid line with squares*) and SURFPRO (*dotted line with triangles*) during summer thunderstorm episode in Torino

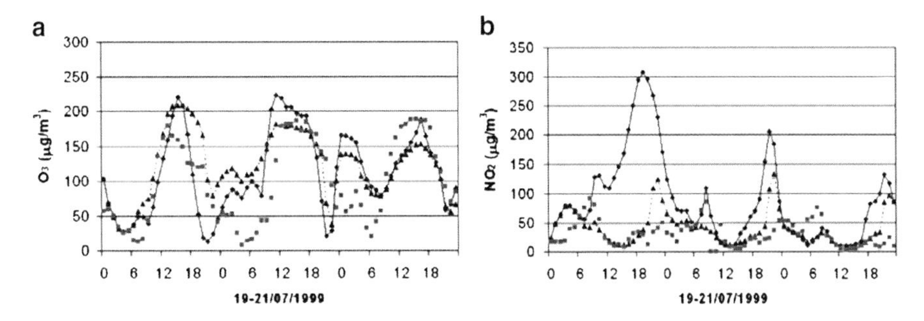

Fig. 9.6 Concentrations of O_3 (**a**) and NO_2 (**b**) computed using RAMS (*dotted line with triangles*) and SURFPRO (*solid line with diamonds*) turbulent fluxes and scaling parameters vs. observations (*grey squares*)

Running the model with the Harrington instead of Chen scheme (Cotton et al. 2003), the NO_X overestimation could be reduced due to larger values of surface radiation obtained.

9.3.4 Air Quality Initialisation at Regional and Urban Scale Effect

During the evaluation of the AQ forecasting system for the Torino metropolitan area (Finardi et al. 2008) the effect on model results of AQ initial and boundary conditions has clearly emerged. To define these data the modelling system relies on CHIMERE continental forecasts provided by *Prev'Air* European Scale Air Quality Service (http://www.prevair.org). This large scale AQ forecast was initially used to define both initial and boundary conditions. In principle local observations could be used to build more realistic initial concentration fields, but they are not yet available at the forecast simulation start time.

The AQ forecasting system results have been compared with observations over a 8 month period: from June 2006 to January 2007. The predicted concentration data have been divided in two time series obtained selecting the first (last) 24 h of each daily forecast cycle that covers the 48 h period. Comparison of these two time series with observations (Fig. 9.7) showed that the 48 h forecast generally obtains higher concentrations and better fits with observations in comparison to the 24 h forecast. This behaviour was common to all pollutants except ozone (which was overestimated for the 24 h simulation). The comparison of initial and 24 h concentration fields showed that differences were due to the influence of initial conditions on the first simulated day. The resolution difference between CHIMERE (50 km) and FARM (4 km) background domains did not allow obtaining a proper initialization. In CHIMERE topography, the city of Torino is located on the slope of the western Alps, at about 800 m asl. This feature clearly favours ozone overestimation and

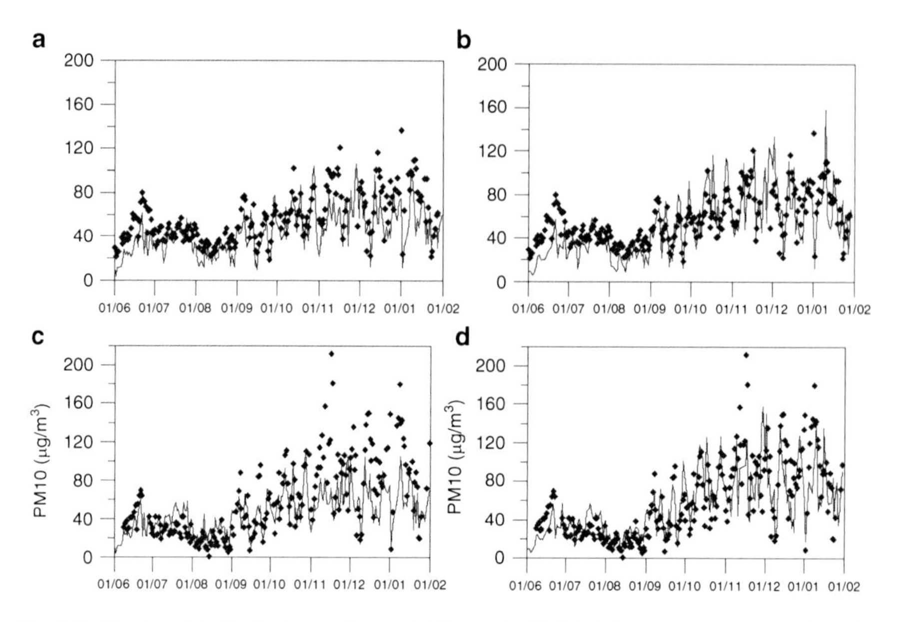

Fig. 9.7 Simulated (*solid line*) and observed (*diamonds*) PM10 daily average concentrations for the (**a**, **c**) first (0:+24) and (**b**, **d**) second (+24:+48) day of forecast at two Torino urban background stations (**a**, **b**) Consolata and (**c**, **d**) Gaidano. The comparison refers to June 2006 – January 2007 time period

underestimation of other pollutants, due to the city location outside of the Po Valley plain.

This identified shortcoming induced changes in the initialization of AQ fields using the forecasting system results of the previous day (+24 h fields) and to prepare a simple AQ analysis tool, to be able to correct initial fields with local observations when they will be available. A resolution match problem is present, even if it is less evident, within the boundary conditions, too. The final set-up of the system is, therefore, programmed to move to three nested domains, introducing a larger background computational mesh enhancing background pollution simulation and reducing the effects of boundary conditions on the AQ forecast. This configuration was originally tested in the EC FP5 FUMAPEX project (Finardi et al. 2008).

9.4 Summary and Discussion

The Working Group 2 of COST-728 compiled a survey of the multiple and diverse coupled meteorological and air quality modelling systems developed and applied covering mainly European countries. The attention was focused on off-line coupled modelling systems, which are by far the more numerous, even if the number of on-line coupled systems recorded is growing fast. The important role of interface modules was discussed and the more common approaches followed in their development were briefly described on the basis of a COST-728 inquiry and the analysis of known modelling systems. In principle, the direct interfacing methods consisting of the evaluation of dispersion parameters from average fields of meteorological models and turbulent fluxes should be preferred to guarantee the modelling system consistency. On the other hand, interfaces/processors allowing the user to reconstruct missing variables and dispersion parameters, in order to increase space resolution and possibly to enhance meteorological fields for lower atmospheric layers and for the urban atmosphere, are widely used by the air quality community. The influence of different interface modules on integrated air quality modelling system results has been shown through application examples taken from the authors' experience.

In the diverse landscape of European modelling, model harmonisation remains an important issue despite earlier efforts, e.g. COST-710 (1994–1998) which are continued in the regular Harmonisation conferences. Modular modelling, flexible input–output strategies and adaptable interfaces following agreed guidelines for off-line and on-line integrated modelling, which are applied by all including the large consortia and community models, would greatly facilitate model improvement and applicability for European users. The large variety of modelling systems can be considered a scientific richness but also highlights problems in the inter-comparison of model results and the collaboration for model development in Europe.

References

Arya SPS (1988) Introduction to micrometeorology. Academic, San Diego, CA, p 307

Baklanov A, Clappier A, Fay B, Joffre S, Karppinen A, Ødegård V, Slørdal LH., Sofiev M, Sokhi RS, Stein A (2005) Improved interfaces and meteorological pre-processors for urban air pollution models. In: Finardi S (ed) Deliverable 5.2-3 of the EC FUMAPEX project, Milan, Italy, 55 pp

Baklanov A, Fay B, Kaminski J, Sokhi R (2007) Overview of existing integrated (off-line and on-line) mesoscale meteorological and chemical transport modelling systems in Europe, Joint Report of COST Action 728 and GURME, GAW Report No. 177, WMO TD No. 1427, May 2008

Beljaars ACM, Holtslag AAM (1991) Flux parametrisation over land surfaces for atmospheric models. J Appl Meteorol 30:327–341

Calori G, Silibello C (2004) FARM (Flexible Air quality Regional Model) Model formulation and user manual. ARIANET Report, Milan, Italy, 59 p

Calori G, Clemente M, De Maria R, Finardi S, Lollobrigida F, Tinarelli G (2005) Air quality integrated modelling in Turin urban area. Environ Model Softw 21(4):468–476

Cotton WR, Pielke RA, Walko RL, Liston GE, Tremback CJ, Jiang H, McAnelly RL, Harrington JY, Nicholls ME, Carrio GG, McFadden JP (2003) RAMS 2001: current status and future directions. Meteorol Atmos Phys 82:5–29

Finardi S, De Maria R, D'Allura A, Cascone C, Calori G, Lollobrigida F (2008) A deterministic air quality forecasting system for Torino urban area, Italy. Environ Model Softw 23:344–355

Gariazzo C, Silibello C, Finardi S, Radice P, Piersanti A, Calori G, Cecinato A, Perrino C, Nussio F, Pellicciori A, Gobbi GP, Di Filippo P (2007) A gas/aerosol air pollutants study over the urban area of Rome using a comprehensive chemical transport model. Atmos Environ 41:7286–7303

Hanna SR, Chang JC (1992) Boundary-layer parametrizations for applied dispersion modelling over urban areas. Bound Layer Meteorol 58:229–259

Mahrt L (1999) Stratified atmospheric boundary layers. Bound Layer Meteorol 90:375–396

Pielke RA, Cotton WR, Walko RL, Tremback CJ, Lyons WA, Grasso LD, Nicholls ME, Moran MD, Wesley DA, Lee TJ, Copeland JH (1992) A comprehensive meteorological modelling system – RAMS. Meteorol Atmos Phys 49:69–91

van Ulden AP, Holtslag AAM (1985) Estimation of atmospheric boundary layer parameters for diffusion applications. J Clim Appl Meteorol 24:1196–1207

Zilitinkevich SS, Perov VL, King JC (2002) Near-surface turbulent fluxes in stable stratification: calculation techniques for use in general-circulation models. Q J R Meteorol Soc 128:1571–1587

Chapter 10
Coupling Global Atmospheric Chemistry Transport Models to ECMWF Integrated Forecasts System for Forecast and Data Assimilation Within GEMS

Johannes Flemming, A. Dethof, P. Moinat, C. Ordóñez, V.-H. Peuch, A. Segers, M. Schultz, O. Stein, and M. van Weele

10.1 Coupling of Earth-System Components Models

Numerical models simulating specific components and aspects of the earth-system have to exchange their results in order to study the many interactions within the earth-system.

A coupled model *A* provides more detailed information about processes which have been treated by simpler assumptions such as explicit or implicit climatologies in model *B*. If the system is two-way coupled, the response in model *B* is fed back to model *A*, leading to different result there, which may again influence model *B*.

Besides the scientific questions related to the coupling of models, the interaction of the numerical models is a big technical challenge. The transformation of data at different temporal and spatial resolutions as well as computational efficiency, memory consumption, data storage capacity, meta-data communication and code management are issues which have to be addressed.

The most common type of coupling is 2D in space, i.e. the coupled models cover separate 3D domains, such as atmosphere and ocean, which are connected to each other by a 2D interface. Less common is 3D coupling in which both models cover the same or an overlapping spatial domain, e.g. the atmosphere, but consider different aspects of it as in the case of weather forecasts models and chemistry transport models. The amount of data to be exchanged is bigger in 3D coupling and there are further consistency issues if both models simulate the same processes such as transport but in a different way (see Section 10.2.4).

There are various options for the technical implementation of the coupling, which differ in the following aspects whether or not:

J. Flemming (✉)
European Centre for Medium Range Weather Forecast, Shinfield Park, Reading, RG2 9AX, UK
e-mail: johannes.flemming@ecmwf.int

A. Baklanov et al. (eds.), *Integrated Systems of Meso-Meteorological and Chemical Transport Models*, DOI 10.1007/978-3-642-13980-2_10,
© Springer-Verlag Berlin Heidelberg 2011

- A dedicated coupling software is used to facilitate the coupling.
- The coupled models stay independent as executables.
- The models or modules run concurrent or step-wise sequential.

The tightest way of coupling, often called "on-line" or "integrated" coupling, is the exchange by argument passing from subroutines simulating different components and aspects of the earth-system. These "integrated" models tends to have less consistency problems because the integrated modules have been aligned to the general model structure, in particular to model geometry and to decomposition for parallelisation. The integration may require a large coding effort, depending on the code structure of the modules, as well as substantial scientific testing to ensure the scientific integrity of the new modules in the existing model. Further, the integrated approach is less flexible in the choice of the coupled model and also requires continuous code management in order to benefit from further development of the included models.

For these reasons many coupled systems try to keep some sort of independence of the component models, and the coupling is facilitated by a specific coupling-software. Ford and Riley (2002) give an overview of coupler software developed in North America and Europe.

The coupling-software mainly consists of two interconnected entities:

- A mechanism to let coupled models runs together and enable them to exchange data
- An infrastructure to process meta-data needed for the communication of the models

The model developer who wants to couple the models (1) has to include coupler-specific interfaces in the models, and (2) has to provide the model and coupling meta-data according to standards of the coupling software. Both tasks can be time consuming, and the gain of interoperability has to be balanced against the costs of the implementation.

There are two basic design concepts for the coupling software: "Concurrent coupling" means that independent model executables or modules run at the same time on different computer resources. An additional coupler/driver executable controls the data exchange between the models. "Sequential coupling" relies on a "superstructure" which calls the components models sequentially for each coupling-time step using the same system resources. Sequential couplers can be considered as a partly automated procedure for the "manual" integration of component models as subroutines in one unified model code. Figure 10.1 shows the schematic of a concurrently coupled system (the GRG system) and integrated/sequential – coupled system. Sequential coupling tends to have less latency problems than concurrent coupling if the two components differ in their computation time. OASIS4 (Valcke and Redler 2006) and OASIS3 (Valcke 2006) are examples of concurrent couplers. The Earth System Modelling Framework (ESMF; http://www.esmf.ucar.edu) is an example for "sequential" coupling.

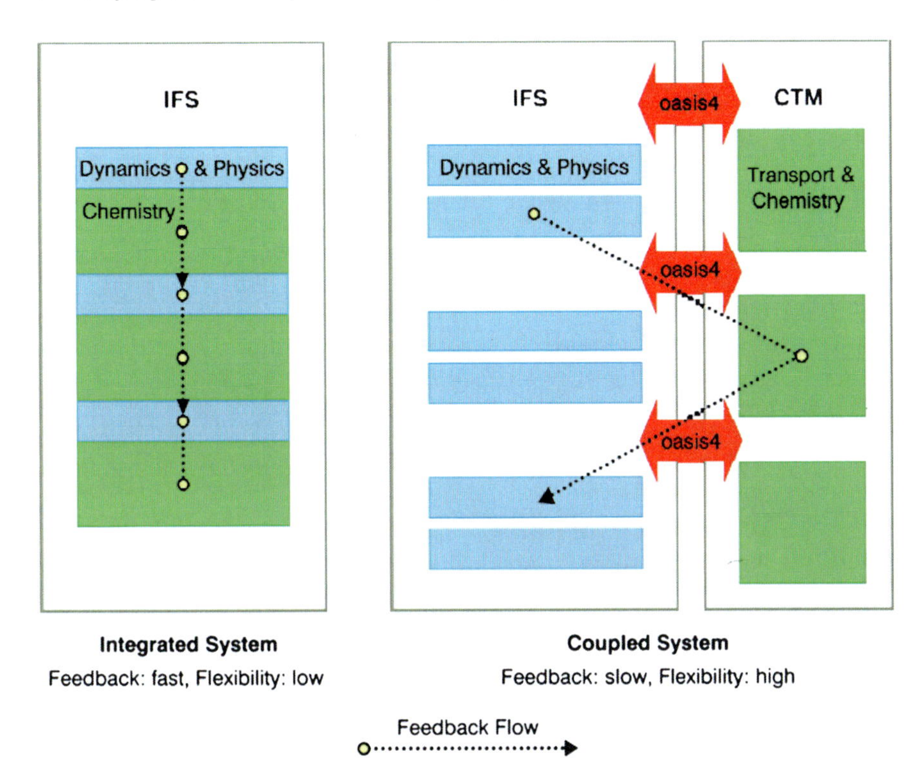

Fig. 10.1 Different designs of a coupled system: Concurrent coupling of two independent executables (*left*) and sequential coupling of model or integration of sub-routine calls (*right*)

10.2 The GEMS GRG Coupled System at ECMWF

As part of the GEMS subproject on global reactive gases (GRG), a system coupling ECMWFs integrated forecast system (IFS) with atmospheric chemical transport models (CTMs) has been developed. The coupler software OASIS4, which is being developed as part of the prism project and the prism support initiative (http://www.prism.enes.org/), is used to couple the two components in concurrent fashion. The OASIS4 interfaces in the IFS and the run and configuration environment for the coupled experiments can easily be adapted to couple other earth system model to the IFS.

The main motivation for the development of the GRG coupled system was the need to account for sink and source processes, namely chemical conversion, in the assimilation of satellite observation of chemical tracers within the 4D-VAR data assimilation of the IFS. Since it appeared too costly to be costly to integrate complex chemical mechanisms in the IFS, tendencies describing sink and source processes should be provided by well established CTMs, being coupled to the IFS. The three CTMs for the GRG coupled system are MOCAGE (Josse et al., 2004), MOZART (Horowitz et al., 2003) and TM5 (Krol et al., 2005).

A clear benefit of the coupled approach, in contrast to integration, is the flexibility in the choice of different coupled chemical schemes represented by the different CTMs. However, the required 3D coupling is less consistent than an integrated system because of feedback delay and dislocation due to different transport representations (see Sect. 10.2.4) in the IFS and the CTM.

10.2.1 Configuration of the GRG Coupled System

The GRG coupled system is a 3D two-way coupled system: IFS provides atmospheric fields at high temporal resolution to drive the CTMs, and the IFS receives tracer concentration fields and tracer tendencies due to source and sink processes from the CTM. A further coupling option is the feedback of concentration fields from IFS to the CTM.

Depending on which tendency data are exchanged, the GRG coupled system can be run in three modes:

- CTM forecast mode
- IFS tracer forecast mode
- IFS tracer data assimilation mode

The CTM forecast mode is a one-way coupling in which IFS provides the meteorological data on-line to the CTM. The main difference to CTM off-line runs is the high temporal resolution at which the CTM gets the atmospheric data. Typical frequency for the coupling is 1 h whereas the temporal frequency in off-line runs is 3–6 h.

In IFS tracer forecast mode the CTM provides initial condition for the chemical tracers (NO_x, NO_2, SO_2, CO, HCHO and O_3) and 3D fields of tracer tendencies due to emissions, deposition and chemical conversion to IFS. The IFS simulates the horizontal and vertical transport of these tracers and applies the CTM tendency data in order to account for the source and sink processes not simulated in the IFS. The CTM itself run as in CTM forecast mode. The feedback option enables replacing the CTM concentration fields, in particular the initial conditions, with the tracer fields of the IFS.

In IFS tracer data assimilation mode, the IFS tracer forecast mode is applied in the outer loops of ECMWF data assimilation system, i.e. the calculation of the trajectories runs of the "complete" model of the 4D VAR (Mahfouf and Rabier 2000) The inner loops used in the minimisation step with the tangent linear and adjoint model are currently run uncoupled, i.e. without the application of the source and sink tendencies from the CTM.

In the coupled system the IFS runs in a T159 spectral resolution and the grid point space is represented in the reduced Gaussian grid (Hortal and Simmons 1991). The vertical coordinate system is given by 60 hybrid sigma-pressure levels. In order

Table 10.1 Resources of the IFS and the CTMs MOZART-3, MOCAGE and TM5

Component	Resolution	Time step (s)	Species	MPI/openMP	Run time 24 h
IFS	T159, 60L	1,800	5	8/2	(Stand alone) 3 min
MOZART-3	T63, 60L	900	106	8/8	12 min
MOCAGE	$2^\circ \times 2^\circ$, 60L	900	126	1/12	188 min
TM5	$2^\circ \times 3^\circ$, 60L	1,200	54	12/1	31 min

to avoid difficulties in the vertical interpolation by the coupler, the CTM use the same 60 vertical levels. The coupler only has to perform horizontal interpolations for which the bi-linear mode is applied. The resolution of the CTM is lower (~T63) as the IFS resolution because of the high computational cost of the CTMs (see Table 10.1).

The IFS is run on a higher horizontal resolution because of the quality of the meteorological forecasts and because a lower resolution would limit the acceptance of high resolution observations within data assimilation. The coupling frequency is 3,600 s which is the largest acceptable time step for the IFS at a T159 resolution, and also the time step of some of the CTMs. The exchange of data can be either provided via a master-process only or via a direct exchange via all processes involved in the simulations.

The following variables are covered by the OASIS4 interfaces in the IFS and the CTMs:

- IFS to CTM
  - T, Q, U, V (3D grid point)
  - O_3, NO_x, SO_2, CO, HCHO – concentration (3D grid point)
  - ps, taux, tauy, shflx, qflx (2D grid point)
  - Vorticity, Divergence, ps (3D/2D spectral fields)
  - Wavenumber-info (3D/2D spectral fields)
- CTM to IFS (3D grid point)
  - O_3, NO_x, SO_2, CO and HCHO tendencies due to chemistry, wet deposition and atmospheric emissions
  - O_3, NO_x, SO_2, CO and HCHO tendencies due to surface fluxes (emission, dry deposition)
  - O_3, NO_x, SO_2, CO and HCHO – concentration

10.2.2 Initial Condition Handling Within Coupled Experiments and Feedback

The coupled long-term simulation and data assimilation runs are structured as a sequence of coupled runs (6 h in data assimilation, 24 h in forecast mode) because the IFS needs a re-start from a meteorological analysis as often as possible. The CTM provides the tracer initial conditions of the IFS for the first forecasts. There are three modes of how the initial conditions for the GRG-tracers are obtained in the subsequent forecasts (see Fig. 10.2).

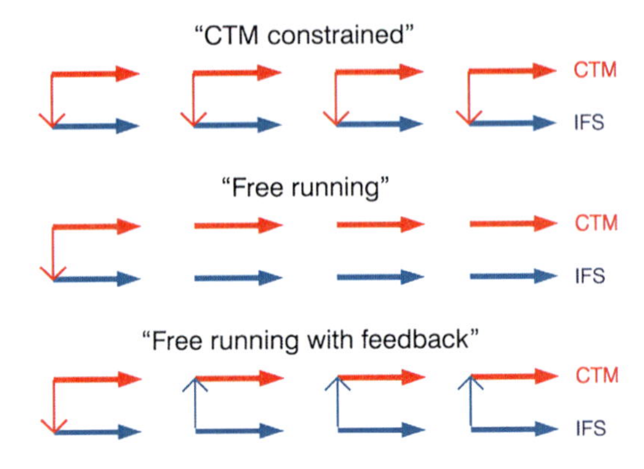

Fig. 10.2 Modes of initial condition handling in a sequence of short simulations

In the "CTM constrained" mode, the IFS gets the initial tracer conditions from the CTM at the start of each forecast. The CTM gets the whole set of initial conditions from the previous CTM run.

In "feedback" mode, the CTM will use the tracer initial conditions provided by IFS after the first model run. The IFS tracer fields may now contain information from observations (analysis mode) or may be different from the CTM fields because of different vertical transport simulation in IFS.

10.2.3 Computational Performance of the GRG Coupled System

The main factor for the computational performance of the coupled GRG-system is the individual run time of the IFS and the CTMs at ECMWF high performance computing facility (IBM power5). The computational cost of the CTMs is clearly higher than the one of the IFS in forecast mode (see Table 10.1). The good scalability of the MOZART-3 model at ECMWFs computer led to acceptable run time within the coupled system. However, the MOZART-3 run time is still three times longer than the one of the IFS using only 25% of the CPUs. Further improvements in the run time of TM5 and MOCAGE are required to achieve acceptable run time within the coupled system.

The overhead because of the coupling can be attributed to the couplers set-up phase (only once per run) and the time of the data transfer and interpolation at every coupling time step. In the given setup the overhead is below about 3% of the IFS stand-alone run time and about 1% of the overall run time with coupled system IFS-MOZART-3.

A further constrained is the memory consumption of the component models and the OASIS4 coupler. The memory usage of the coupler occurs only temporarily during the exchange events but can reach up to 60% of the IFS memory consumption, 15% of the MOZART-3 and 12% of the total consumption. Figure 10.3 shows the memory consumption for each mpi-process of the coupled system IFS-MOZART-3.

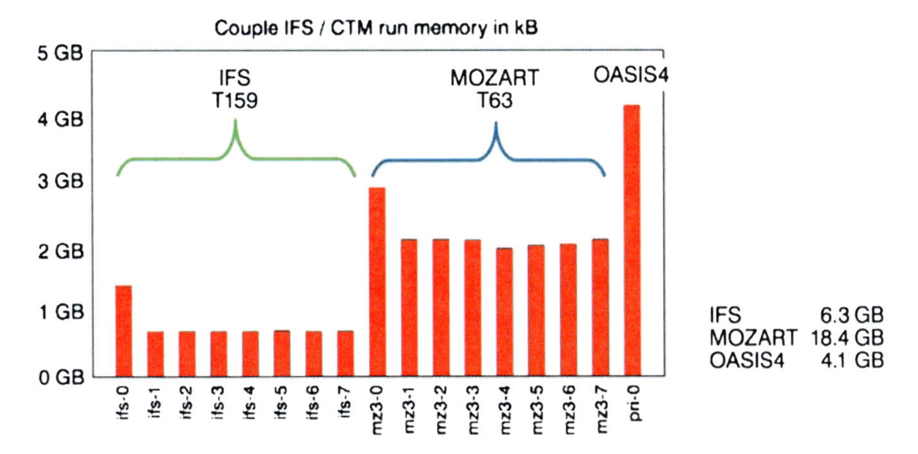

Fig. 10.3 Memory consumption of the MPI-tasks of the OASIS4 coupled system IFS – MOZART

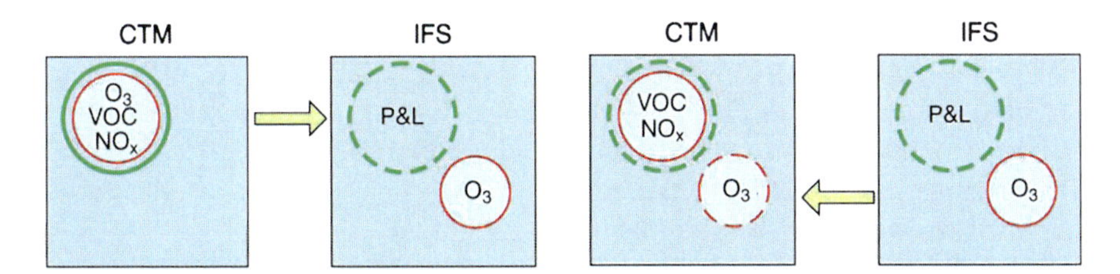

Fig. 10.4 Dislocation problem: a mismatch in the ozone fields (O_3, red) between CTM and IFS causes a mismatch in the application of ozone tendency data (P&L, green) transferred from CTM to IFS

10.2.4 Dislocation and Feed-Back Delay

In the case of the GRG coupled system, both the IFS and the CTM simulate atmospheric transport processes. Different advection schemes or spatial and temporal resolutions may lead to different concentration fields in the IFS and the CTM. Thus, the applied CTM tendencies can be inconsistent with the concentration fields in the IFS. The most annoying consequence would be negative concentration values in the IFS, due to un-balanced loss processes.

One example of the dislocation problem is depicted in Fig. 10.4. O_3 tendency data due to chemical conversion (P&L, green circle) shall be given from a CTM to IFS which does not simulate chemistry. If the O_3 fields in CTM and IFS are dislocated, the tendencies data will be applied in the wrong part of the model domain.

Two-way coupling is required if one wants to study the feedback of processes not included in the models. However, the time scales for the interaction is limited at least by two times the coupling interval. Two-way coupling requires a synchronous run of the two models. Lagged two-coupling, in which one of the component models runs ahead of the other model, is possible if the first model is not sensitive to delayed input from the model running behind. Lagged two-way coupling can be an option in atmosphere-ocean coupling but it was no option for the GRG-system.

In contrast to one-way coupling or lagged two-way coupling, the information for the next time step is not available in two-way coupling. This makes impossible to forward-extrapolate the external data, e.g. meteorological fields, in time. Instead, they have to be assumed to be constant over the coupling interval.

10.3 Specific Issues of the GRG Coupled System

10.3.1 Formulation of Tendency Terms

The exchange of concentration tendencies, rather than concentrations, is a special and perhaps unique feature of the GRG coupled system. The formulation of the tendency terms has to reflect the operator splitting and time stepping in the both the CTMs and the IFS as well as the relation between the tendency and the respective concentration value, and the cost (memory, time) of the exchange.

The CTM use an operator-splitting approach in which chemistry, emission injection, diffusion and deposition are called in sequence and the update of the concentration follows directly within each subroutine.

The total tendencies T is given by the sum of chemical loss L_C and production P_C, gain due to emissions P_E and loss L_E due to deposition.

$$T = P_C - L_C + P_E - L_D$$

Deposition L_D and chemical loss L_C and are generally proportional to the tracer concentration x and a relative formulation $L = l\,x$, i.e. a loss rate l, would better link tendency and concentration value and would help to avoid negative concentration. However, the output arguments of chemical routines provide total tendencies $(P_C + L_C)$ for each time step and it would be difficult to distinguish production and loss. The relative formulation of the production is not advisable because it could cause high concentrations values to become even higher. One option would be to link the loss to OH concentrations only.

A disadvantage of separating production and loss, which tend to be much larger in absolute values than the resulting total, seems to be the separate interpolation of these fields. The sum of the interpolated production and loss terms may suffer from im-balances close to strong gradient, in particular if non-linear interpolation is applied.

Emissions are independent of the tracer concentration and can be considered as a surface flux. The injection of the emissions is integral part of the diffusion scheme in MOZART-3, i.e. as lower boundary for the fluxes, whereas TM5 and MOCAGE distribute the injected mass in a fixed ratio over selected layers in the boundary layer and apply their diffusion operator after the injection. The tendencies of the emissions P_E, therefore, have to be formulated either as 3D field including the diffusion or as 2D flux term. The diffusion in the IFS would have to be switched off if the 3D emissions-diffusion tendencies are applied. Air born emissions such as the ones from aircraft would have to be included in the 3D chemistry tendencies, if the surface emissions are expressed as a flux.

Dry deposition occurs at the lowest level and is expressed both as tendency for the lowest layer or as a flux. Wet deposition would be a 3D tendency field.

The consideration of the arguments discussed above led to the following implementation of tendency extraction within the CTM:

- Process-specific 3D tendencies are determined by calculating the difference of the concentration fields before and after each of the chemistry, emission/diffusion and deposition subroutines.
- The process-specific 3D tendencies are averaged over the coupling interval.
- The process-specific tendencies are either added up to *one* 3D total tendency field or added up to *two* 3D tendency fields containing (1) chemistry and wet deposition and (2) emission and dry deposition.
- The one or two 3D tendency fields are transferred to the IFS.

Depending on a control switch, the 3D emission and dry deposition tendencies can be converted into a surface flux by calculating the total column integral within the IFS.

Figure 10.5 shows profiles of the tendencies due to chemistry and wet deposition as well as emissions including vertical diffusion and convection for NO_x and CO at 12 and 24 UTC. The data is area-averaged over Central Europe (42N/-10W – 55N/10E) and shown in units of kg/m^{2}s to demonstrate the mass contribution of each model level. Model levels 60-50 cover the PBL, the tropopause is about at level 30. Clearly visible is the day–night difference of chemical loss and production. The emissions are a constant source term but the vertical tendency profiles are shaped by the vertical exchange in the boundary layer.

10.3.2 Implementation of GRG-Tracers Tendency Application in the IFS

The simplified sequence of the simulation of a passive tracer within an IFS time-step is as follows:

- Calculate tendencies due to semi-Lagrangian advection scheme
- Calculate tendencies due to "physics":
  - Calculate tendencies due to surface flux injection and vertical diffusion within one routine

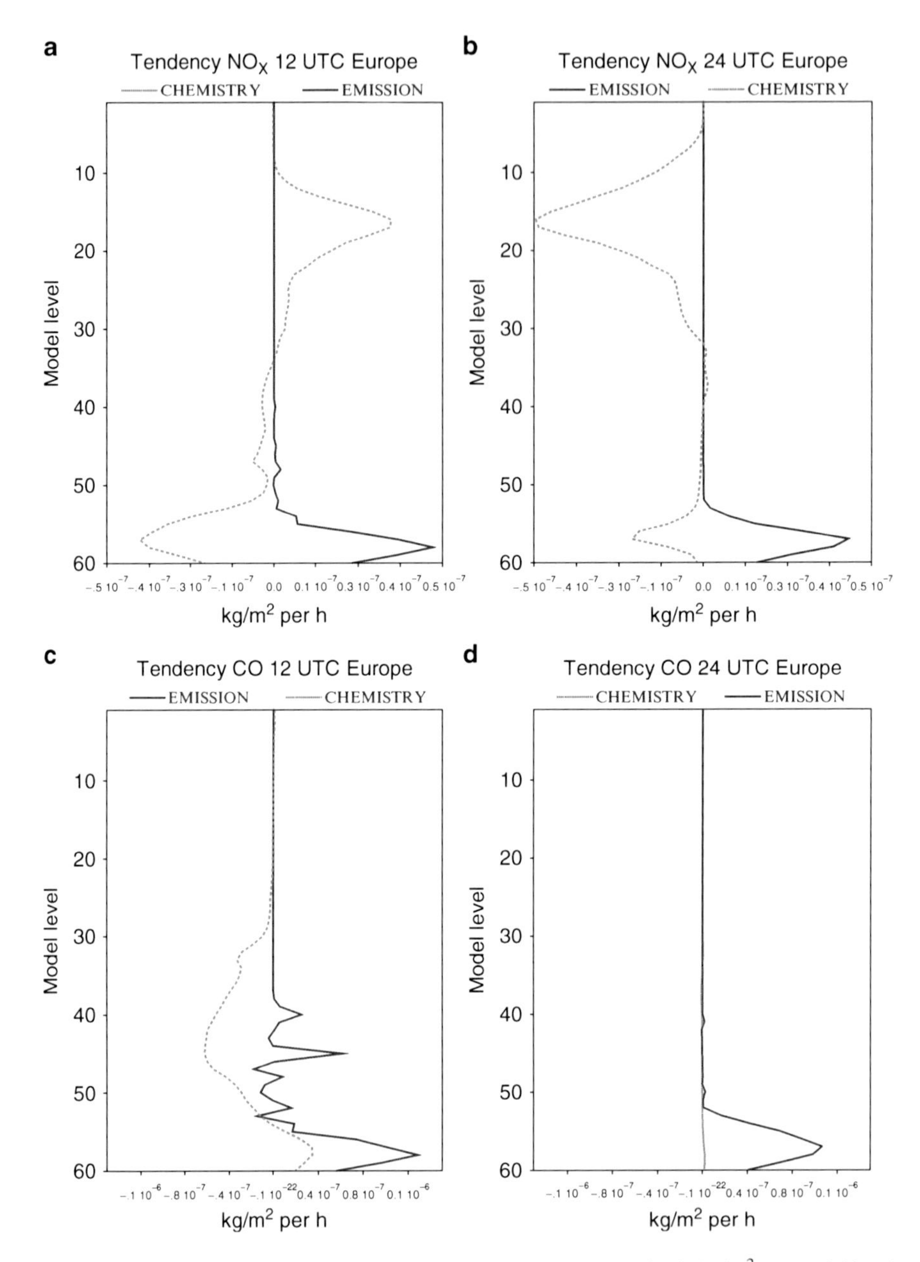

Fig. 10.5 Profile of the area averaged (over selected domain) tendencies in kg/m^2 per model level due chemistry and emissions including vertical diffusion and convection for (**a**, **b**) NO_x and (**c**, **d**) CO at 12 and 24 UTCs

  - Calculate tendencies due to convection
  - Calculate tendencies due to "other" processes (e.g. chemistry parametrization)
- Update concentration fields from the start with accumulated tendencies of advection and "physics"

Although the process-specific tendencies are stored for the update in a final step, the processes are not treated in an independent, i.e. parallel way. This is because the diffusion-routine uses the concentrations updated with dynamical tendencies and the convection routine uses the concentration updated with the diffusion (and surface flux) tendencies.

The position of the processes in the time loop is influenced by whether the process is fast or slow with respect to the time scale. More details on the implementation of the IFS physics can be found in Beljaars et al. (2004). Since the coupling interval (1–3 h) is larger than the model time step, the processes parameterised by CTM input will appear as slow processes, even if the actual chemical conversion can be rather quick in the CTMs.

In the IFS, emission injection and diffusion are part of one subroutine. Surface emissions, and likewise (dry) deposition, can be treated as surface fluxes. If the applied CTM tendencies already included the effect of diffusion and convection, the respective routine in the IFS physics would have to be switched off for this GRG-tracer.

The CTM source and sink information can be implemented in two modes:

- IFS with complete CTM "physics" for tracers: All "physics" tendencies (diffusion, convection, emission, chemical conversion, deposition) come from the CTM.
- IFS with CTM chemistry tendencies (3D) and CTM surface fluxes (emission and dry deposition), diffusion and convection within the IFS.

In the first mode, the IFS would only advect the GRG-tracers. The CTM tendency field, consisting of the contributions of all source and sink processes would be consistent in itself. Dislocation can occur due to different advection in the CTM and the IFS.

In the second mode, a consistent treatment of the emission injection and vertical transport would be achieved. In particular, the adjoint formulation of diffusion and convection in data assimilation would be consistent with the forward model. However, dislocation of the chemistry tendencies is more likely than in case 1 because the IFS concentration fields tend to differ more from the CTM fields.

10.3.3 A Diagnostic NO_x Inter-conversion Operator for Fast Reaction Not Captured by the Coupled Approach

The fast diurnal NO_2 – NO inter-conversion caused by solar radiation in the upper stratosphere could not be handled by the coupled system with a coupling frequency of 1 h. Instead of a steady movement of the day–night border, a "carved" stripe-shaped concentration fields were simulated. Therefore, it was decided to use NO_x as the model variable since the chemical development of the NO_x field is not so strongly influenced by solar radiation and the development of the NO_x field can be simulated by the coupled system.

Since the satellite observations to be assimilated are NO_2 data, a diagnostic NO_x to NO_2 inter-conversion operator H was developed. For the application in 4D VAR data assimilation it's tangent linear **H** and adjoint $\mathbf{H}^T$ had to be coded.

The inter-conversion operator is based on a simple chemical equilibrium between the NO_2 photolysis rate j_{NO_2} and the O_3 concentration:

$$\frac{[NO_2]}{[NO_x]} \approx \frac{K_2[O_3]}{j_{NO_2} + k_2[O_3]}$$

The diagnostic NO_2/NO_x ratio depends on the following variables:

- Solar zenith angle
- O_3 concentration
- Slant O_3 column above
- Temperature

A parameterised approach for the calculation of clear-sky NO_2 – photolysis j_{NO_2} rates was used based on the band scheme by Landgraf and Crutzen (1998) in combination with actinic fluxes parameterised following Krol and Van Weele (1997). The diagnostic operator does not reflect the influence of clouds on j_{NO_2}, and the adjustments to the equilibrium because of hydro-carbons in the lower troposphere and abundant O-radical in the higher stratosphere and mesosphere.

The missing cloud influence might be tolerable in the assimilation system since the NO_2 observations tend to be restricted to conditions with small cloud cover (Boersma et al. 2004).

The inter-conversion operator links the NO_2 to the O_3 concentration in data assimilation. It is the first step towards the consideration of more chemical relationships within the GRG data assimilation system.

Figure 10.6 shows a profile of the NO_2/NO_x ratio over Europe at 12 UTC calculated by the diagnostic operator and directly from the MOZART NO and NO_2 fields. An ad-hock approach of assuming a per-oxy-radical ($HO_2 + RO_2$) concentration of 80 ppt (Kleinman et al. 1995) in the troposphere, multiplied by the cosine of the solar zenith angle to account of the diurnal cycle of the in the per-oxy-radical concentration, improved further the match of the NO_2/NO_x ratio between the diagnostic operator and MOZART.

Photolysis frequencies in MOZART-3 are based on tabulated values of the Tropospheric Ultraviolet and Visible radiation model ((TUV) version 3.0) (Madronich and Flocke 1998) for clear sky conditions. The adjustment for cloudiness is described in Brasseur et al. (1998).

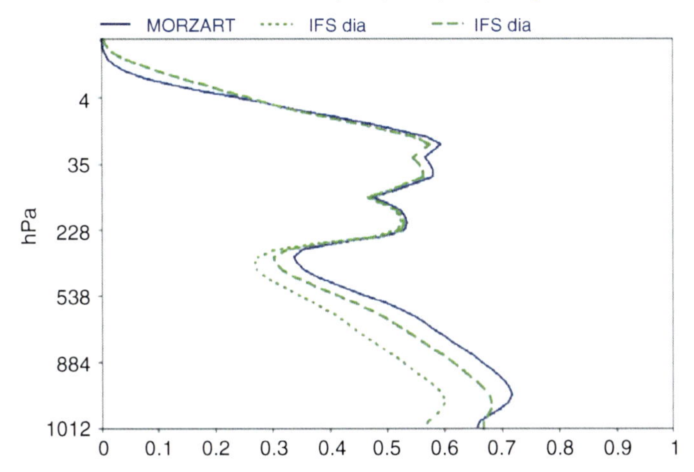

Fig. 10.6 NO_2–NO_x ratio profile averaged over Europe on 07 Jan 2002, 12 UTC, taken from MOZART directly and from the diagnostic inter-conversion operator without (IFS_dia dotted) and with and ad-hoc assumption of tropospheric peroxy-radical concentration (IFS_dia dashed)

10.4 Testing the Scientific Integrity of the GRG Coupled System in Forecast Mode

The integrity of the coupled system depends on whether the application of external tendency fields accounting for processes not included in IFS (chemistry, emission and deposition) give reasonable results of the forecast length. The objective is that the IFS is able to imitate the CTM concentration developments and does not produce to many negative concentrations due to dislocated loss processes.

We studied area-averaged time series of tracer concentrations and spatial patterns of concentrations fields. The following IFS runs are compared with the MOZART concentrations:

- IFS_free: Initial conditions from MOZART, IFS transport
- IFS_tend: Initial conditions from MOZART, IFS advection and CTM sink & source tendencies including vertical transport
- IFS_chem: Initial conditions from MOZART, IFS advection, diffusion and convection CTM sink and source tendencies excluding vertical transport

Figure 10.7 shows examples of time series of the area average over Europe of the GRG species for model level 55 (PBL niveau) of the three IFS runs and the MOZART simulation. If total CTM tendencies (IFS_tend) are applied, the IFS can imitate the CTM up to a forecast length of 48 h. Differences are obvious if the IFS vertical transport scheme is applied, because the vertical transport schemes

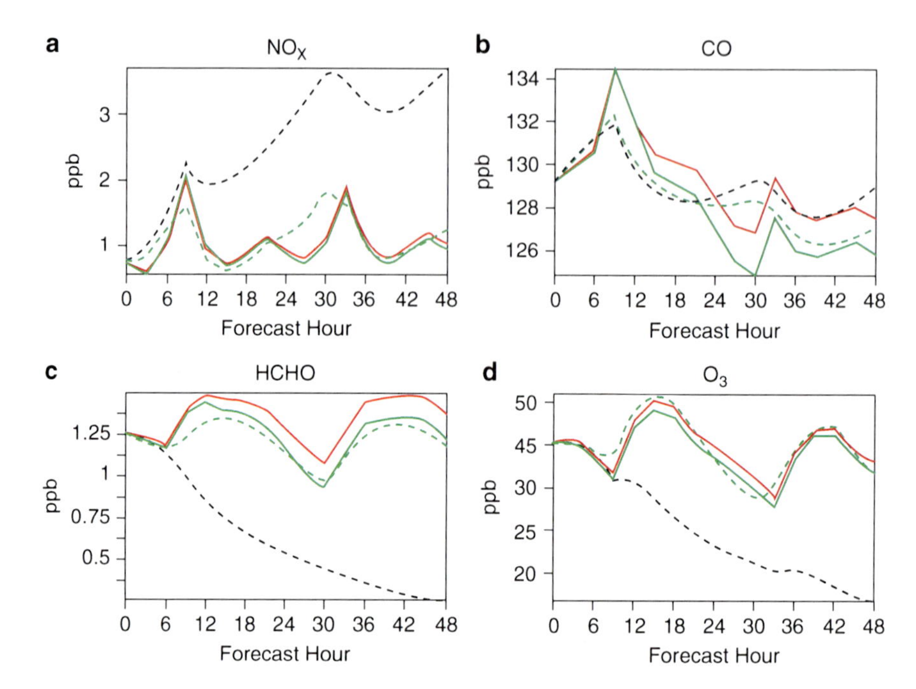

Fig. 10.7 Time series of area mean of pollutant NO_x, CO, HCHO, and O_3 concentration over selected European domain simulated with the MOZART, with IFS using no tendency information (IFS_free), with IFS using CTM source and sink tendency information including diffusion and convection (IFS_tend) and with IFS using CTM source and sink tendency and IFS vertical transport and emission injection (IFS_chem)

differ between MOZART and the IFS. Spurious negative NO_x concentrations were detected during the night time, when the IFS vertical transport was applied (IFS_chem).

References

Beljaars A, Bechtold P, Kohler M, Morcrette J-J, Tompkins A, Viterbo P, Wedi N (2004) The numerics of physical parameterization. Seminar on Recent developments in numerical methods for atmospheric and ocean modelling. 6–10 Sep 2004. (http://www.ecmwf.int/publications/library/do/references/show?id=86397

Boersma KF, Eskes HJ, Brinksma EJ (2004) Error analysis for tropospheric NO_2 retrieval from space. J Geophys Res 109:D04311, doi:10.1029/2003JD003962

Brasseur GP, Hauglustaine DA, Walters S, Rasch PJ, Müller J-F, Granier C, Tie XX (1998) MOZART, a global chemical transport model for ozone and related chemical tracers: 1. Model description. J Geophys Res 103:28265–28289

Ford RW, Riley GD (2002) FLUME coupling review, UK met-office. http://www.metoffice.gov.uk/research/interproj/flume/pdf/d3_r8.pdf

Horowitz LW et al (2003) A global simulation of tropospheric ozone and related tracers: description and evaluation of MOZART, version 2. J Geophys Res doi:10.1029/2002JD002853
Hortal M, Simmons AJ (1991) Use of reduced Gaussian grids in spectral models. Mon Weather Rev 119:1057–1074
Josse B, Simon P, Peuch V-H (2004) Rn-222 global simulations with the multiscale CTM MOCAGE. Tellus 56B:339–356
Kleinman L, Lee Y-N, Springston SR, Lee JH, Nunnermacker L, Weinstein-Lloyd J, Zhou X, Newman L (1995) Peroxy radical concentration and ozone formation rate at a rural site in the southeastern United States. J Geophys Res 100(D4):7263–7274
Krol M, Van Weele M (1997) Implications of variation of photodissociation rates for global atmospheric chemistry. Atmos Environ 31:1257–1273
Krol MC, Houweling S, Bregman B, van den Broek M, Segers A, van Velthoven P, Peters W, Dentener F, Bergamaschi P (2005) The two-way nested global chemistry-transport zoom model TM5: algorithm and applications. Atmos Chem Phys 5:417–432
Landgraf J, Crutzen PJ (1998) An efficient method for online calculations of photolysis and heating rates. J Atmos Sci 55:863–878
Madronich S, Flocke S (1998) The role of solar radiation in atmospheric chemistry. In: Boule P (ed) Handbook of environmental chemistry. Springer, Heidelberg, pp 1–26
Mahfouf JF, Rabier F (2000) The ECMWF operational implementation of four-dimensional variational assimilation. Part I: experimental results with improved physics. Q J R Meteorol Soc 126:1171–1190
Valcke S, Redler R (2006) OASIS4 user guide (OASIS4_0_2). PRISM Support Initiative Report No 4, 64 pp
Valcke S (2006) OASIS3 user guide (prism_2-5). PRISM support initiative Report No 3, 64 p

Chapter 11
The PRISM Support Initiative, COSMOS and OASIS4

René Redler, Sophie Valcke, and Helmuth Haak

11.1 The PRISM Concept, Goals, and Organization

PRISM (http://prism.enes.org/) was initially started as a project under the European Union's Framework Programme 5 (FP5, 2001–2004) and its long term support is now ensured by multi-institute funding of seven partners (CERFACS, France; NEC Europe Ltd., Germany; CGAM, UK; CNRS, France; MPI-M&D, Germany; Met Office, UK; and ECMWF) and nine associate partners (CSC, Finland; IPSL, France; Météo-France, France; MPI-M, Germany; SMHI, Sweden; and computer manufacturers CRAY, NEC-HPCE, SGI, and SUN) contributing with IT and Earth science experts to the PRISM Support Initiative (PSI), currently for a total of about 8 person-years per year.

The PRISM concept, initially a recommendation made by the Euroclivar initiative (Anderson et al. 1998), is to increase what Earth system modellers have in common today (compilers, message passing libraries, algebra libraries, etc.) and share also the development, maintenance and support of a wider set of ESM software tools and standards. This should reduce the technical development efforts of each individual research team, facilitate the assembling, running, monitoring, and post-processing of ESMs based on state-of-the-art component models developed in the different climate research centres in Europe and elsewhere, and therefore, promote the key scientific diversity of the climate modelling community. As demonstrated in other fields, sharing software tools is also a powerful incentive for increased scientific collaboration. It also stimulates computer manufacturers to contribute, thereby increasing the tool portability and the optimization of next generation of compute server for ESM needs, and also facilitating computer manufacturer procurement and benchmarking activities. The extensive use of the OASIS coupler illustrates the benefits of a successful shared software infrastructure. In 1991, CERFACS was commissioned to realise specific software for

R. Redler (✉)
NEC Laboratories Europe – IT Divison, NEC Europe Ltd., Sankt Augustin, Germany
e-mail: rene.redler@zmaw.de

A. Baklanov et al. (eds.), *Integrated Systems of Meso-Meteorological and Chemical Transport Models*, DOI 10.1007/978-3-642-13980-2_11,
© Springer-Verlag Berlin Heidelberg 2011

coupling different geophysical component models developed independently by different research groups. The OASIS development team strongly focussed on efficient user support and constant integration of the developments fed back by the users. This interaction snowballed and resulted in a constantly growing community. Today, the OASIS development and support capitalises about 32 person-years (py) of mutual developments and fulfils the coupling needs of about 25 climate research groups around the world.

PRISM represents the first major collective effort at the European level to develop ESM supporting software in a shared and coherent way. This effort is recognised by the Joint Scientific Committee (JSC) and the Modelling Panel of the World Climate Research Programme (WCRP) that has endorsed it as a "key European infrastructure project". It is analogous to the ESMF project (http://www.earthsystemmodeling.org) in the United States.

PRISM is lead by the PRISM Steering Board (one member per partner) that reviews each year a work plan proposed by the PRISM Core Group composed of PSI Coordinator(s), the leaders of the PRISM Areas of Expertise (see the next paragraph), and the chair of the PRISM User Group. The PRISM User Group is composed of all climate modelling groups using the PRISM software tools; given the dissemination of the OASIS coupler, the PRISM User Group is already a large international group.

11.1.1 PRISM Areas of Expertise

PRISM is organised around five PRISM Areas of Expertise (PAEs) having the following remits:

- Promote and, if needed, develop software tools for ESM. A tool must be portable, usable independently and interoperable with the other PRISM tools, and freely available for research. There should be documented interest from the community to use the tool and the tool developers must be ready to provide user support
- Encourage and organise a related network of experts, including technology watch
- Promote and participate in the definition of community standards where needed
- Coordinate with other PRISM areas of expertise and related international activities

11.1.1.1 PAE "Code Coupling and I/O"

The scope of the PAE "Code coupling and I/O" is to:

- Develop, maintain, and support tools for coupling climate modelling component codes
- Ensure a constant technology watch on coupling tools developed outside PRISM
- Keep strong relations with the different projects involving code coupling in climate modelling internationally

The current objectives are to maintain and support the OASIS3 coupler, continue the development of the OASIS4 coupler (Redler et al., 2010), but also, through the organisation of workshops, help the community understand the different technical approaches used in code coupling, for example in the PALM coupler (Buis et al. 2006), in the UK Met Office FLUME project (Ford and Riley 2003), in the US ESMF project (Killeen et al. 2006), and in the Bespoke Framework Generator (BFG) from U. of Manchester (Ford et al. 2006) used in the GENIE project. Currently OASIS3 is used for coupling between the COSMOS components.

11.1.1.2 PAE "Integration and Modelling Environments"

This PAE targets the following environments:

- Source version control for software development (including model development)
- Code extraction and compilation
- Job configuration (how to set up and define a coupled integration)
- Job running (how to control the execution of a coupled integration)
- Integration with archive systems

For source version control, PRISM promotes the use of subversion (Pilato et al. 2004) and recently moved from CVS to Subversion for its own software distribution server, now located at DKRZ in Hamburg.

Controlling the creation of executables and providing a suitable run time environment were seen as key integration activities within the EU FP5 PRISM project, resulting in the development of the PRISM Compile and Running Environments (SCE & SRE). These tools are built on the concept that software from multiple sources can use a single framework as long as those models can conform to a set of simple standards.

Different groups (ECMWF, IPSL, CERFACS) have also shown strong interest in the UK Met Office Flexible Configuration Management (FCM) tool for version control and/or compilation management. A further review will be conducted in those groups and this tool, together with Subversion, may be considered as a replacement to CVS and extend the simplicity of 'make' in the current SCE. The COSMOS components follow these standards and are integrated into the PRISM SCE and SRE.

PrepIFS is a flexible User Interface framework provided by ECMWF that allows tailored graphical user interfaces to be built for the configuration of models and other software. It is integrated within the Supervisor Monitor Scheduler (SMS) for the management of networks of jobs across a number of platforms and both products are developed using Web Services technology. SMS and prepIFS have recently been packaged for use within the Chinese climate community. The power of these tools is recognised by PRISM, even if they are not widely used in the European climate community because of the level of commitment and human resources required to run these sophisticated services.

CERFACS is currently developing a Graphical User Interface to configure a coupled model using the OASIS4 coupler.

11.1.1.3 PAE "Data Processing, Visualisation and Management"

The overall objective of this PAE is the development of standards and infrastructure for data processing, archiving, and exchange in ESM and more general Earth System Research (ESR). The huge amounts of data in ESR do not allow for centralised data archiving. Networking between geographically distributed archives is required. Ideally the geographical distribution of federated data archives is hidden to the user by an integrative graphical WWW-based interface. Standards are required in order to establish a data federation and to work in a network.

For data processing, this PAE currently analyses the Climate Data Analysis Tools (CDAT) and Climate Data Operators (CDO) respectively maintained and developed by PCMDI and MPI-M.

The M&D Group also develops the CERA-2 data model for the World Climate Data Centre, proposing a description of geo-referenced climate data (model output) and containing information for the detection, browse and use of data. An important collaboration is going on with the PAE Metadata and other international initiatives for the development and implementation of metadata standards for the description of model configuration and numerical grids.

Collaboration between PRISM related data archives in the development of data networking and federated archive architectures is also going on. ECMWF MARS software may be another candidate tool for meteorological data access and manipulation even if it is likely to be of interest only to major NWP sites due to its complexity.

11.1.1.4 PAE "Metadata"

In the last few years metadata has become a hot topic with new schemes and ideas to promote the interchangeability of ESMs or modelling components as well as data. The PRISM Metadata PAE provides a forum to discuss, develop, and coordinate metadata issues with other national and international projects. The fundamental objective is to develop, document, and disseminate ESM metadata schemes as well as tools for producing, checking and displaying metadata. Currently, this PAE offers an opportunity to ensure coherence between the following metadata definition efforts:

- Numerical Model Metadata (NMM; http://ncas-cms.nerc.ac.uk/NMM), developed at University of Reading, is an evolving international metadata standard intended for the exchange of information about numerical code bases, and the models/simulations done using them.

- The CURATOR project (Earth System Curator; http://www.earthsystemcurator.org), a project similar to NMM in the US.
- Numerical grid metadata, developed by Balaji at GFDL, USA, for numerical grid description (http://www.gfdl.noaa.gov/~vb/gridstd/gridstd.html).
- The CF convention for climate and forecast metadata Interface (Eaton et al. 2003) designed to promote the processing and sharing of files created with the netCDF Application Programmer, developed by an international team and now managed by PCMDI and BADC.
- The metadata defined by the OASIS4 developers for description and configuration of the coupling and IO interface in a coupled model.
- The metadata currently defined in the UK Met Office FLUME project to manage and define the process of model configuration.

The goal of this PAE is therefore to integrate these emerging standards, ensure that they meet requirements and needs of the ESM community, and disseminate them as a "good practice".

11.1.1.5 PAE "Computing"

Experience has shown that a large variety of technical aspects related to computing are highly important for Earth system modelling. These techniques are in constant flow and evolve with new hardware becoming available. While computer vendors have to be kept informed about requirements emerging from the climate modelling community, Earth system modellers still have to be informed about computing issues to preview difficulties and evolutions. PRISM can play a role in that aspect through the new PAE "Computing" devoted to those technology trends. Possible technical topics are file IO and data storage, algorithmic development, portable software to fit the needs of parallel and vector systems.

It is first proposed to establish a group of people willing to contribute with their expertise via mailing list and a sharing of relevant information on the PRISM web site. In particular the activities will cover sharing of knowledge from the work on the Earth Simulator, establishment of links with the DEISA project (http://www.deisa.eu), and providing information about important conferences and workshops. Depending on the number of volunteers and the input from this group, the list of tasks will be revised or extended in the next years.

11.2 COSMOS: COmmunity Earth System MOdelS

The idea behind the *CO*mmunity Earth *S*ystem *MO*del*S* (COSMOS) is that a single research institution alone cannot develop the most comprehensive models. COSMOS is a community network towards the development of a fully developed ESM. The complexity of ESMs requires the involvement of an interdisciplinary team of scientists to develop ESMs. This team has to be ready to work together intensively and share the knowledge and expertise of the team partners. So, COSMOS constitutes a team of experts to develop a flexible and portable model infrastructure, following and supporting the ideas of the PRISM initiative. The purpose is not only to develop models and their infrastructure, but also to use them to address challenging problems involving the interactions between different components of the Earth system. These models will be central tools to assess important feedback processes in the system, to assess environmental risks, and to develop mitigation and adaptation strategies.

11.2.1 Organisation

The most important body of the COSMOS network – community of Earth system scientists – is ready to work with scientific tools and endeavours to promote the software. The community meets, at least, once a year for the COSMOS General Assembly. At these meetings it takes strategic decisions, e.g. new scientific co-operation and projects. It also determines the configuration of the board (12 members). The members need to come from institutions having signed the COSMOS Memorandum of Understanding. Three of these members are one Chair and two Co-chairs; the board decides about these positions. The board calls activity heads for working groups and other tasks. The activity heads assignment is to organise and run groups to think about and act upon projects, methods, aspects, etc. of ESM. The activity heads report about the activities of their enterprises to both the board and community. In their daily business they interact intensively with the COSMOS office.

The COSMOS office consists of the Science Director, an ex officio member of the board, and a Project Manager running day-to-day business. The office gets secretarial assistance for office issues and meeting organisation. Support positions for technical work on the model system as well as outreach activities are applied for. The office supports the chair and the co-chairs. It is assigned by the board, and it develops and proposes the strategies and plans for COSMOS.

11.2.2 The COSMOS Models

The COSMOS network crucially depends upon the availability and easy access to models. These models do not only have to be very well tested and mature, but also scientifically and technologically very advanced. The MPI-M has developed a suite of models in a framework, called "COSMOS v1" (http://cosmos.enes.org/). The configurations possible with this modelling framework are indicated in Table 11.1.

This is available to the COSMOS network. It is assembled following the PRISM philosophy:

Table 11.1 Model configurations with COSMOS v1

	Model components
Atmosphere GCM	ECHAM5
Ocean GCM	MPIOM
Ocean + biogeochemistry	MPIOM-HAMOCC
Atmosphere/ocean GCM	ECHAM5/OASIS3/MPIOM
Carbon cycle	ECHAM5-JSBACH/OASIS3/MPIOM-HAMOCC
Aerosol system	ECHAM5-HAM/OASIS3/MPIOM-HAMOCC

- Coupling of atmosphere and ocean GCMs by OASIS3
- Configuration of model types and building of executables for specific machines by the standard configuration environment (PRISM SCE)
- Running and data storage by using the standard runtime environment (PRISM SRE) where model configurations can be ported to, run, maintained and developed on supercomputers with relative ease

This first version of the model of the COSMOS network, called COSMOS v1, has 28 users from 10 countries.

11.2.3 Atmosphere Chemistry

The chemistry of the atmosphere can be simulated with the new MESSy approach, the Modular Earth Submodel System (http://www.messy-interface.org). It has been successfully coupled to ECHAM5, and applied in multiyear integrations in different configurations and resolutions. ECHAM5/MESSy can be used to simulate ozone and related chemistry of the lower and middle atmosphere up to the mesopause at about 80 km, with no artificial boundaries applied, e.g., between the troposphere and stratosphere. The modular structure allows the selection of particular configurations to increase or decrease the level of details in describing processes such as tropospheric multiphase chemistry, aerosols, transport and deposition. The COSMOS v1 package integrates in a flexible and modular way models for the circulation of the atmosphere, ocean and sea ice, and optionally includes processes for aerosols, vegetation, and marine biogeochemistry. The integration of atmospheric chemistry into the COSMOS system will lead to the next generation: COSMOS v2. This will take profit from the ECHAM5 based atmospheric chemistry models:

- ECHAM5-MESSy for tropospheric and stratospheric chemistry (already distributed)
- ECHAM5-HAMMOZ for coupled aerosols and chemistry in the troposphere
- HAMMONIA for the neutral and ionized chemistry, covering the entire atmosphere

The PRISM standard environments of COSMOS v2 will allow for different model configurations, including atmospheric chemistry.

11.3 The OASIS Coupler

The OASIS coupler software allows synchronized exchanges of coupling information between numerical codes representing different components of the climate system.

OASIS3 (Valcke 2006) is the direct evolution of previous versions of the OASIS coupler. In addition a new fully parallel coupler OASIS4 (Valcke and Redler 2006) is developed within PRISM. Other MPI-based parallel coupling software performing field transformation exists, such as the Mesh based parallel Code Coupling (MpCCI; http://www.mpcci.de) or the NCAR CCSM Coupler 6 (Cpl6; http://www.cesm.ucar.edu/models/ccsm3.0/cpl6/). The originality of OASIS in general lies in its great flexibility (as the coupling configuration is externally by the user) and in the common treatment of coupling and I/O exchanges (again externally defined by the user).

As the climate modelling community is progressively targeting higher resolution climate simulations on massively parallel platforms with coupling exchanges involving a higher number of (possibly 3D) coupling fields at higher coupling frequencies, a completely new fully parallel coupler OASIS4 is developed within PRISM. OASIS4 is a portable set of Fortran 90 and C routines. At run-time OASIS4 acts as a separate parallel executable, the OASIS4 Driver-Transformer, and as a fully parallel model interface library, the OASIS4 PSMILe. The concepts of parallelism and efficiency like the parallel neighbourhood search drove OASIS4 developments, keeping at the same time in its design the concepts of portability and flexibility that made the success of OASIS3 and its predecessors.

11.3.1 Coupling Configuration

Each component model to be coupled via OASIS4 should be released with an eXtensible Markup Language (XML; http://www.w3.org/XML) file describing all its potential input and output fields, i.e. the fields that can be received or sent by the component through PSMILe *get* and *put* actions in the code. Based on those description files, the user produces, either manually or via a Graphical User Interface, the XML configuration files. As for OASIS3, the OASIS4 Driver extracts the configuration information at the beginning of the run and sends it to the different model PSMILes, which then perform the appropriate coupling or I/O actions during the run. OASIS4 is also highly flexible in the sense that any duration of run, any number of component models, any number of coupling and I/O fields, and particular coupling or I/O parameters for each field, can be specified.

11.3.2 Process Management

In a coupled run using OASIS4, the component models remain separate executables. If only MPI1 is available (Snir et al. 1998), the OASIS4 main processes (driver plus transformer routines) and the component models must all be started at once in the job script. If some or all of the components are programmed as subroutines of a main program (e.g. ocean and sea ice as subroutines of ocean general circulation model) these components can still be coupled via OASIS4 PSMILe routines provided that the subroutines are run concurrently. This can usually be achieved by running the subprograms on different processes which is configurable through the XML file and the use of appropriate PSMILe interface routines. If the MPI library supports the MPI2 standard (Gropp et al. 1998) the user has the option to start only the OASIS4 driver process which then launches the different component models and transformer processes using the *MPI_Comm_spawn_multiple* functionality. The OASIS4 driver can spawn the different processes on different machines. In both cases, all processes are necessarily integrated from the beginning to the end of the run, and each coupling field is exchanged at a fixed frequency defined in the XML file for the whole run. In that sense, OASIS4 supports static coupling only.

Figure 11.1 illustrates the different ways of communication between the individual physical components with a coupled application. Ocean and sea ice are programmed as subroutines of a main program. In this example we assume that the two grids are identical and data can be exchanged directly through the PSMILe bypassing the transformer processes. Atmosphere and chemistry are programmed as separate executables. As they again work on identical grids data are exchanged directly (indicated by the lower red double arrow). As those grids are different from the ocean and sea ice data exchanged with those components go through the parallel transformer.

11.3.3 Coupling Field Transformation and Regridding

During the run the OASIS4 Driver (root) process takes over the functionality of the transformer and together with the other Transformer processes manages the transformation and regridding of 2D or 3D coupling fields. The (parallel) Transformer performs only the weight calculation and the regridding per se; the neighbourhood search, i.e. the search of the source points determination for each target point that contribute to the calculation of its regridded value, is performed in parallel in the source PSMILe.

During the simulation time stepping, the OASIS4 parallel Transformer can be considered as an automaton that reacts to what is demanded by the different component model PSMILes: receive data for transformation (source component process) or send transformed data (target component process). The OASIS4 Transformer, therefore, acts as a parallel buffer in which the transformations take place. Currently, only 2D and 3D nearest-neighbour, 2D and 3D linear, and bi-cubic regridding, and 2D conservative remapping techniques are implemented, but there are plans to implement also 3D cubic grid interpolation and 3D conservative remapping.

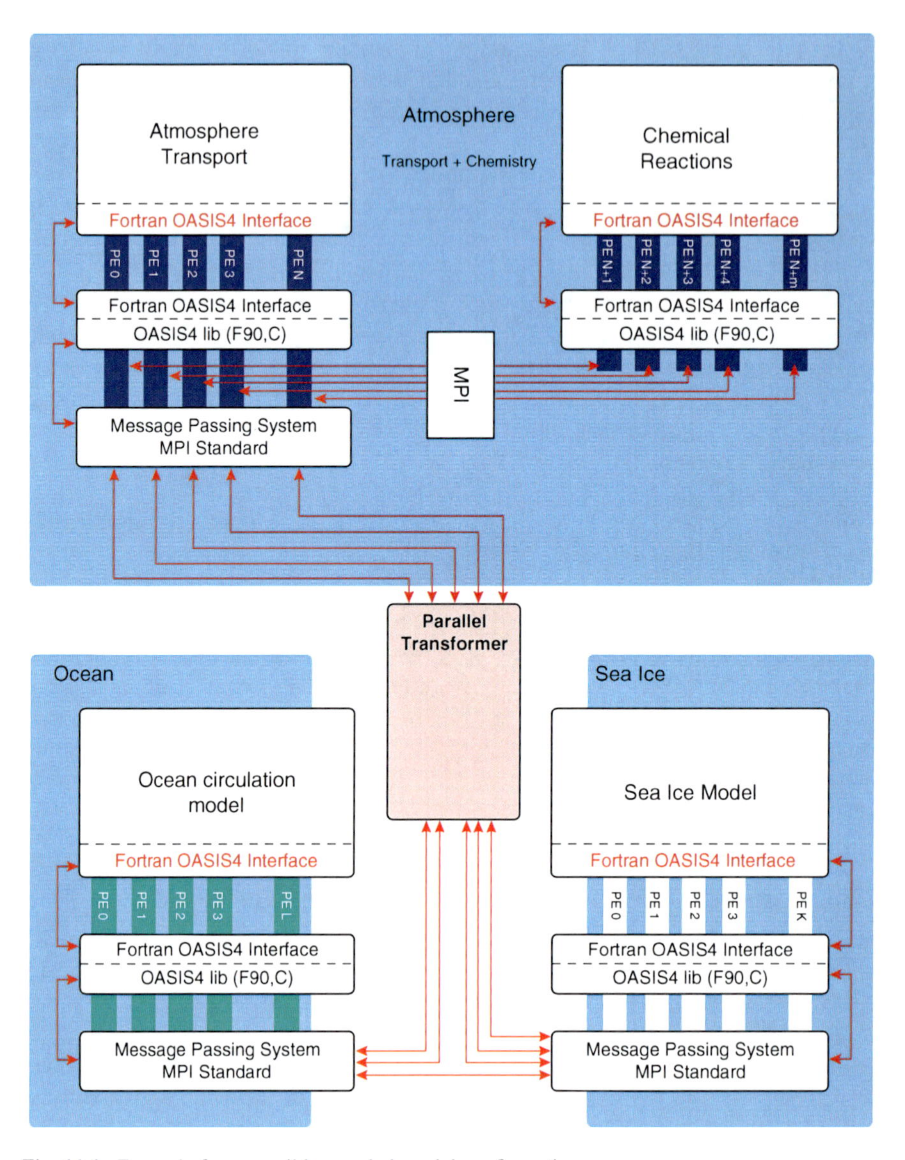

Fig. 11.1 Example for a possible coupled model configuration

11.3.4 Communication: The OASIS4 PSMILe Software Layer

To be coupled via OASIS4, the component models have to include specific calls to the OASIS4 PSMILe software layer. The OASIS4 PSMILe Application Programming Interface (API) was kept as close as possible to OASIS3 PSMILe

API; this ensures a smooth and progressive transition between OASIS3 and OASIS4.

The OASIS4 PSMILe supports fully parallel MPI-based communication, either directly between the models for those pairs of source and target grid points for which an exact match was found (including automatic repartitioning if needed) or via the parallel Transformer, and file I/O using the GFDL mpp_io library. Note, the mpp_io library has been extended to work optionally with the parallel NetCDF library (pNetCDF; http://cucis.ece.northwestern.edu/projects/PNETCDF) to allow for parallel file I/O via MPI-IO inside pNetCDF. The detailed communication pattern among the different component model processes is established by the PSMILe, using the results of the regridding or repartitioning neighbourhood search. This search is based on the source and target identified for each coupling exchange by the user in the XML configuration files and on the local domain covered by each component process. The search uses an efficient multigrid algorithm and is done in parallel in the source PSMILe, which ensures that only the useful part of the coupling field is extracted and transferred.

Besides these new parallel aspects, the OASIS4 PSMILe follows the same end-point communication and user-defined external configuration principles than the OASIS3 PSMILe.

11.3.5 The OASIS4 Users

OASIS4 portability and scalability was demonstrated with different "toy" models during the EU FP5 PRISM project. OASIS4 was also used to realize a coupling between the MOM4 ocean model and a pseudo atmosphere model at the Geophysical Fluid Dynamic Laboratory (GFDL) in Princeton (USA), and with pseudo models to interpolate data onto high resolution grids at IFM-GEOMAR in Kiel, Germany.

Currently, work is going on with OASIS4 at the:

- Swedish Meteorological and Hydrological Institute (SMHI) in Sweden for coupling regional ocean and atmosphere models (first physical case studies are already realized)
- European Centre for Medium-Range Weather Forecast (ECMWF), KNMI in the Netherlands, and Météo-France in the framework of the EU GEMS project, for 3D coupling between atmospheric and chemistry models
- UK MetOffice for global ocean-atmosphere coupling
- CERFACS and Météo-France
- ACCESS

After the current beta-testing phase, the first official OASIS4 version will become available to the public in 2008.

11.4 Final Remarks

PRISM is already a success as it allows a community of developers facing similar technical problems in ESM to share their expertise and ideas. The difference in the level of buy-in for the different tools developed during the FP5 project helped identified for which tool standardisation is more or less achieved, for which tools convergence is wanted, and is currently not a target. The strength of the current decentralised PRISM organisation is to allow "best of breed" software tools to naturally emerge, although this means that PRISM relies on the developments done in the different partner groups to propose technical software solution to Earth system modellers. In the areas for which this philosophy does not apply and for which standards have to be pre-defined, for example for metadata definition, the big contribution of PRISM is to provide a visible entry point of the European ESM software community for international coordination, for example with the American ESMF project within the WCRP framework. Given its institutional long term support, PRISM is now well placed to seek additional funding to support more networking and coordination activities or to help specific technical developments. And of course, more European or non-European collaborators are today most welcome to bring in additional expertise and to ensure a wider diffusion of the PAE tools and standards.

Appendix A: List of Acronyms

ACCESS	Australian community climate earth-system simulator
CDAT	Climate data analysis tool
CDO	Climate data operators
CERA	Climate and environmental retrieval and archive
CF	Climate and forecast
COSMOS	Community earth system models
CVS	Concurrent versions system
DEISA	Distributed European infrastructure for supercomputing applications
ESM	Earth system model
ESR	Earth system research
EU	European Union
FCM	Flexible configuration management
FP5	Framework Programme 5
GCM	General circulation model
GFDL	Geophysical fluid dynamics laboratory
IO	Input and output
JSC	Joint scientific committee
MARS	Meteorological archival and retrieval system
MPI1	Message passing interface Version 1
MPI2	Message passing interface Version 2
MPI-M	Max-Planck-Institute for meteorology
MpCCI	Mesh-based code coupling interface

(*continued*)

NetCDF	Network common data form
NMM	Numerical model metadata
NWP	Numerical weather prediction
OASIS	Ocean atmosphere sea ice soil
PAE	PRISM area of expertise
pNetCDF	Parallel network common data form
PRISM	Program for integrated earth system modelling
PSI	PRISM support initiative
PSMILe	PRISM system model interface library
SCE	Standard compile environment
SMS	Supervisor monitor scheduler
SRE	Standard run environment
WCRP	World climate research programme
XML	eXtensible Markup Language

References

Anderson D, Bengtsson L, Delecluse P, Duplessy J-C, Fichefet T, Joussaume S, Jouzel J, Komen G, Latif M, Laursen L, Le Treut H, Mitchell J, Navarra A, Plamer T, Planton S, Ruiz de Elvira A, Schott F, Slingo J, Willebrand J (1998) Climate variability and predictability research in Europe, 1999–2004: Euroclivar recommendations. KNMI, De Bilt, The Netherlands, 120 pp

Buis S, Piacentini A, Déclat D (2006) PALM: a computational framework for assembling high performance computing applications. Concurr Comput: Practice and Experience 18(2): 247–262. http://www.cerfacs.fr/globc/PALM_WEB/

Eaton B, Gregory J, Drach B, Taylor K, Hankin S (2003) NetCDF Climate and Forecast (CF) Metadata Conventions, Version 1.0. http://www.cgd.ucar.edu/cms/eaton/cf-metadata/CF-1.0.html

Ford RW, Riley GD (2003) The met office FLUME project – high level design. Manchester Informatics Ltd., The University of Manchester, 15 pp. http://www.cs.manchester.ac.uk/cnc/projects/met.php

Ford RW, Riley GD, Bane MK, Armstrong CW, Freeman TL (2006) GCF: a general coupling framework. Concurr Comput: Practice and Experience, 18 (2):163–181. http://www.cs.manchester.ac.uk/cnc/projects/bfg.php

Gropp W, Huss-Lederman S, Lumsdaine A, Lusk E, Nitzberg B, Saphir W, Snir M (1998) MPI – the complete reference, vol 2, The MPI extensions. MIT, Cambridge, MA

Killeen T, DeLuca C, Gombosi T, Toth G, Stout Q, Goodrich C, Sussman A, Hesse M (2006) Integrated frameworks for Earth and space weather simulation. American Meteorological Society Meeting, Atlanta, GA. http://www.earthsystemmodeling.org

Pilato CM, Collins-Sussman B, Fitzpatrick BW (2004) Version control with subversion, O'Reilly Media Inc., 320 pp. http://svnbook.red-bean.com/

Redler R, Valcke S, Ritzdorf H (2010) OASIS4 â€ a coupling software for next generation earth system modelling. Geosci Model Dev 3:87–104

Snir M, Otto S, Huss-Lederman S, Walker D, Dongarra J (1998) MPI – the complete reference, vol 1, The MPI core. MIT, Cambridge, MA

Valcke S (2006) OASIS3 user guide (prism_2-5). PRISM Support InitiativeTechnical Report No 3, 64 pp. (http://prism.enes.org/Publications/Reports/oasis3_UserGuide_T3.pdf)

Valcke S, Redler R (2006) OASIS4 user guide. PRISM Support InitiativeTechnical Report No 4, 72 pp. http://prism.enes.org/Publications/Reports/OASIS4_User_Guide_T4.pdf

Chapter 12
Integrated Modelling Systems in Australia

Peter Manins, M.E. Cope, P.J. Hurley, S.H. Lee, W. Lilley, A.K. Luhar, J.L. McGregor, J.A. Noonan, and W.L. Physick

12.1 Introduction

The Commonwealth Scientific and Industrial Research Organisation (CSIRO) is Australia's principal research organisation with over 6,000 scientific, technical and support staff at 65 sites around Australia. A small group of scientists, in the Division of Marine and Atmospheric Research, has been working on air pollution models, developing, documenting, disseminating, and applying them to environmental issues around Australia and abroad. An extensive literature in both international journals and reports is available (see http://www.cmar.csiro.au/search/pubsearch.htm for author or keyword entry to these).

Motivated to explore air pollution in complex geographic settings, in 1986 Bill Physick coupled the Pielke mesoscale meteorological model (Colorado State University) with McNider's Lagrangian particle model (University of Alabama) in off-line mode. After replacing the meteorological model with one developed locally (by John McGregor), speeding up the particle model near boundaries, and adding a new advanced display system, the result was Lagrangian Atmospheric Dispersion Model (LADM), used in numerous coastal industrial air pollution settings (see Physick et al. (1994) for more discussion of this history and references).

Seeing the strengths and weaknesses of the LADM approach, Peter Hurley built a test bed to explore various turbulence schemes, non-hydrostatic effects, and further ideas for speeding up Lagrangian particle models; by 1999 these had come together on a PC platform as the GUI-driven in-line integrated mixed-Lagrangian/Eulerian modelling system – The Air Pollution Model (TAPM) (Hurley 1999a, b).

P. Manins (✉)
Commonwealth Scientific and Industrial Research Organization (CSIRO), Marine and Atmospheric Research, PMB 1, Aspendale 3195, VIC, Australia
e-mail: peter.manins@csiro.au

A. Baklanov et al. (eds.), *Integrated Systems of Meso-Meteorological and Chemical Transport Models*, DOI 10.1007/978-3-642-13980-2_12,
© Springer-Verlag Berlin Heidelberg 2011

12.2 The Air Pollution Model

TAPM uses the fundamental equations of atmospheric flow, thermodynamics, moisture conservation, turbulence and dispersion, wherever practical. For computational efficiency, it includes a nested approach for meteorology and air pollution, with the pollution grids optionally able to be configured for a sub-region and/or at finer grid spacing than the meteorological grid, which allows a user to zoom-in to a local region of interest quite rapidly. The meteorological component of the model is nested within synoptic-scale analyses/forecasts that drive the model at the boundaries of the outer grid. The coupled approach taken in the model, whereby mean meteorological and turbulence fields are passed to the air pollution module every 5 min, allows pollution modelling to be done accurately during rapidly changing conditions such as those that occur in sea-breeze or frontal situations. The model incorporates explicit cloud microphysical processes. The use of integrated plume rise, Lagrangian particle, building wake, and Eulerian grid modules, allows industrial plumes to be modelled accurately at fine resolution for long simulations. Similarly, the use of a condensed chemistry scheme also allows nitrogen dioxide, ozone, and particulate mass to be modelled for long periods.

TAPM has become an important integrated modelling system in Australia and abroad, evolving to the present Version 4 (Hurley 2008a, b): there are 170 active licences, involving 18 countries. It is successful because: (1) ease to use, and (2) self-imposed limitations that make it very practical:

- Its application is limited to a few thousand kilometers; so, major inter-continental transports are not accommodated
- Because we are focussed on air pollution problems, high-impact weather is rarely of concern; so, its simplified approach to deep convection, absence of a stratosphere (maximum altitude is 8 km), and its explicit but simplified rain processes, are not significant
- It includes wet and dry deposition processes and a simplified photochemical smog mechanism that is only applicable to regional and urban pollution.

The most striking thing about TAPM for most applications by its target user group is that it only requires information on pollutant emissions. I.e. datasets of the important inputs (e.g., terrain, land use and 3D synoptic meteorology every 6 h for target years) are needed for meteorological simulations and provided with the model, allowing model set up for any region, although user-defined databases can be connected to the model if desired. In fact TAPM has been shown to give very good results without any recourse to local meteorological data. Many demonstrations of the veracity of this claim have been made (see Hurley et al. (2008) for an extensive set of case studies), and more recent references are given already at the CMAR Library Web address.

12.3 Learnings from Air Pollution Studies

A major learning for Australian conditions is that recirculation of pollutants in the sea breeze is an important feature of coastal cities: the spatial resolution required of the meteorological model needs to be adequate to resolve the sea breeze phenomenon. Resolution substantially finer than 10 km is required (i.e. 2–5 km seems adequate).

Using a high quality emissions dataset, TAPM performance is very good for the prediction of extreme pollution statistics, important for environmental impact assessments, for both non-reactive (tracer) and reactive (nitrogen dioxide, ozone and particulate) pollutants for a variety of sources (e.g. industrial stacks and surface or urban emissions) – see Luhar and Hurley (2003); Hurley et al. (2005). A study with the attention to detail that makes a big difference, is by Luhar et al. (2006a).

Other findings include:

- Correctly specifying land-use and vegetation improves the meteorological predictions and has flow-through benefits for the air quality predictions
- The vertical temperature profiles in standard synoptic analyses often have poor detail below 1,000 m. For example, extra-stable layers between 200 and 400 m can be non-existent in the forecast model analyses that have only a few levels below 1,000 m. This can greatly affect air quality predictions
- It is important to use wind data only from well-sited anemometers for evaluating model performance, or for assimilating during a run.

Characteristics of urban areas that can affect flow properties include roughness length, building characteristics, thermal properties of the surface and anthropogenic heat flux. As Luhar et al. (2006b) showed by comparison with the Swiss BUBBLE data, although TAPM in particular accounts for these effects with a varying degree of complexity, the land-surface scheme in the model needs improvement to resolve the urban canopy layer and the roughness sublayer. The topic of urbanisation of meteorological models is an area of increasing interest, and has been a focus of COST-728. But there are also important new developments for rough boundary layers in which shedding shear layers slow the flow markedly (Harman and Finnigan 2007). These may have implications for understanding flows in and above urban canopies.

12.4 Larger-Scale Integrated Pollution Modelling in Australia

As TAPM was being developed, the opportunity arose to integrate its chemical transport component with Australian Bureau of Meteorology's weather forecasting system to provide real-time weather and air pollution forecasts. TAPM already had been using historical 6-hourly BoM global analyses to initialise that model; so, the extension was natural. In collaboration with the BoM and the major environment authorities of Australia, we developed the Australian Air Quality Forecasting System (AAQFS) in time to be run operationally for the 2000 Sydney Olympic Games (see Cope et al. (2004), Hess et al. (2004), Tory et al. (2004)). The major emphasis for Sydney was on predicting urban ozone (Cope et al. 2005a). It has been run twice daily ever since, producing hourly forecasts for the next 36 h for 21 chemical species on a 1 km grid for the major cities of Australia. We have learnt a lot and have made many improvements.

- A big learning was that it is best to have as much of the emissions inventory as possible on-line, described by algorithms that respond where appropriate to the forecast meteorology, and calibrated by relevant observational data. We now have emissions from motor vehicles, vegetation (Azzi et al. 2005; Kirstine and Galbally 2004), soils, wind-blown dust, bushfires, sea-salt spray, domestic wood heating, and some industry handled this way.
- As with the finer-scale TAPM experience, it is vitally important to accurately include biogenic emissions. In Australia, concentrations of up to 60 ppb ozone are measured in country areas with no evident industry or vehicle sources.
- Intrusion of bushfire smoke is a major cause of exceedences of air quality standards, in particular for ozone, in Australian cities. This is an extreme example of the effect of biogenic emissions, and is a major driver for the next point.
- Australia-wide forecasts are now done for wind-blown dust (e.g., Wain et al. 2006) and for bushfire smoke. Size-segregated dust is emitted and transported depending on historical land-use, soil-type and seasonal leaf area index (LAI). For bushfires, we use the Sentinel outputs of hotspots to locate emissions. Sentinel currently obtains MODIS data from the NASA EO Satellites Terra and Aqua (see http://sentinel2.ga.gov.au/acres/sentinel).

AAQFS provided the impetus to further explore complex chemistries for air pollution predictions, something we had been doing for special projects in and around Australia (e.g., Cope et al. 2003; Malfroy et al. 2005). We are currently gaining experience with the Carbon Bond 2005 mechanism and have incorporated this into a new Chemical Transport Model that runs optionally in both TAPM and AAQFS (it is called TAPM-CTM). This complexity is essential for addressing policy questions of urban planning and around the veracity of new transport fuels such as ethanol blends, and the effects of pollutants such as formaldehyde and benzene. Pollution by ozone and fine particles, particularly secondary particles, are the main questions ultimately being addressed, though air toxics and personal exposure and the interaction with indoor air pollution are also questions increasingly being asked.

Recognition that personal exposure is the really relevant air pollution question for human health has led to a lot of work on near-road air quality, both experimentally and by modelling. The Lagrangian Wall Model (LWM: Lilley and Cope 2005; Cope et al. 2005b), a complex chemical transport model of a wall that is advected downwind of nominated anchor points can resolve pollution concentrations to 10 m.

The model runs within a TAPM grid cell. Some 30 or so walls can be set up and tracked at once from a GUI, giving high resolution results for the effects of individual roads, intersections and terrain on the air quality. Other relevant work integrates the results of environmental monitoring data, high resolution modelling of pollution fields and hospital admissions to seek to improve understanding of pollutant exposure to increases in asthma (Physick et al. 2006, 2007).

On a wider front, for downscaling from large-scale climate models we have developed the Conformal-Cubic Atmospheric Model (CCAM), a global model that has spatially varying resolution (McGregor 1997; McGregor and Dix 2005). It is initialised from a single global analysis or climate change prediction to predict scenarios of weather down to a kilometre for hours to months. Embedding TAPM off-line and with the possibility of using TAPM-CTM on-line in CCAM gives us a powerful GUI-driven integrated weather and air pollution forecasting system that has wide application.

12.5 Future Directions

Our current direction is to merge our research with Australian Bureau of Meteorology. A particular development underway is ACCESS, the Australian Community Climate and Earth System Simulator, based in large part on HADGEM in collaboration with the UK Hadley Centre (Martin et al. 2004; Johns et al. 2004). CCAM is expected to become an alternative dynamical core (McGregor et al. 2007), and AAQFS will be unified into the system. For air quality applications, ACCESS will play the major role, but TAPM and LWM will continue to develop in parallel, providing very convenient test beds for process and algorithm developments before they are considered for incorporation into ACCESS and possibly HADGEM.

References

Azzi M, Cope ME, Day S, Huber G, Galbally IE, Tibbett A, Halliburton B, Nelson P F, Carras JN (2005) A biogenic volatile organic compounds emission inventory for the Metropolitan Air Quality Study (MAQS) region of NSW [electronic publication]. In: Towards a new agenda: 17th international clean air & environment conference proceedings. Clean Air Society of Australia and New Zealand, Hobart. [Hobart], 7 p. Available: http://www.cmar.csiro.au/e-print/open/cope_2005b.pdf

Cope ME, Hurley PJ, Lilley B, Edwards M, Azzi M, Beer T (2003) The use of a 10% ethanol-blended fuel in the Sydney greater metropolitan region: a modelling study of ozone air quality impacts: final report to Environment Protection Authority of New South Wales CAR:PSS: gy78. CSIRO Atmospheric Research, CSIRO Energy Technology, Aspendale, Vic, 13 p

Cope ME, Hess GD, Lee SH, Tory K, Azzi M, Carras JN, Lilley W, Manins PC, Nelson P, Ng YL, Puri K, Wong N, Walsh S, Young M (2004) The Australian air quality forecasting system. Part I: project description and early outcomes. J Appl Meteorol 43(5):649–662. doi:10.1175/2093.1

Cope ME, Hess GD, Lee SH, Tory KJ, Burgers M, Dewundege P, Johnson M (2005a) The Australian air quality forecasting system: exploring first steps towards determining the limits of predictability for short-term ozone forecasting. Bound Layer Meteorol 116(2):363–384

Cope ME, Lilley W, Marquez L, Smith N, Lee SH (2005b) A high resolution chemically reactive near-field dispersion model. Part 2: Application in an intelligent transport system [electronic publication]. In: Towards a new agenda: 17th international clean air & environment conference proceedings. Clean Air Society of Australia and New Zealand, Hobart. [Hobart], 6 p. Available: http://www.cmar.csiro.au/e-print/open/cope_2005d.pdf

Harman IN, Finnigan JJ (2007) A simple unified theory for flow in the canopy and roughness sublayer. Bound Layer Meteorol 123(2):339–363

Hess GD, Tory K, Cope ME, Lee SH, Puri K, Manins PC, Young M (2004) The Australian air quality forecasting system. Part II: case study of a Sydney 7-day. J Appl Meteorol 43(5): 663–679

Hurley PJ (1999a) The air pollution model (TAPM) Version 1: technical description and examples. CSIRO Atmospheric Research, Aspendale. (CSIRO Atmospheric Research technical paper; no. 43). 41 p. Available: http://www.cmar.csiro.au/e-print/open/hurley_1999a.pdf

Hurley PJ (1999b) The air pollution model (TAPM) Version 1: user manual. CSIRO Atmospheric Research, Aspendale. (CSIRO Atmospheric Research internal paper; 12). 22 p. Available: http://www.cmar.csiro.au/e-print/open/hurley_1999b.pdf

Hurley PJ (2008a) The air pollution model (TAPM) version 4. Part 1. Technical description. CSIRO Atmospheric Research, Aspendale, Vic. Available: http://www.cmar.csiro.au// research/tapm/index.html

Hurley PJ (2008b) The air pollution model (TAPM) version 4. User manual. CSIRO Atmospheric Research, Aspendale, Vic. Available: http://www.cmar.csiro.au/research/tapm/index.html

Hurley PJ, Edwards M, Physick WL, Luhar AK (2005). TAPM V3: model description and verification. Clean Air Environ Qual 39(4):32–36. Available: http://www.cmar.csiro.au/ e-print/internal/hurleypj_x2005d.pdf

Hurley PJ, Edwards M, Luhar AK (2008) The air pollution model (TAPM) version 4. Part 2. Summary of Some Verification Studies. CSIRO Atmospheric Research, Aspendale, Vic. Available: http://www.cmar.csiro.au/research/index.html

Johns T et al (2004) HadGEM1 – Model description and analysis of preliminary experiments for the IPCC Fourth Assessment Report. Hadley Centre Technical Note 55, Sep 2004. Available: http://www.metoffice.gov.uk/research/hadleycentre/pubs/HCTN/index.html

Kirstine W, Galbally IE (2004) A simple model for estimating emissions of volatile organic compounds from grass and cut grass in urban airsheds and its application to two Australian cities. J Air Waste Manag Assoc 54(10):1299–1311

Lilley W, Cope ME (2005) Development of a high resolution chemically reactive near-field dispersion model for assessing the impact of motor vehicle emissions. Part 1: Model description and validation. [electronic publication]. In: Towards a new agenda: 17th international clean air & environment conference proceedings. Clean Air Society of Australia and New Zealand, Hobart. [Hobart], 7 p

Luhar AK, Hurley PJ (2003) Evaluation of TAPM, a prognostic meteorological and air pollution model, using urban and rural point-source data. Atmos Environ 37(20):2795–2810

Luhar AK, Galbally IE, Keywood MD (2006a) Modelling PM10 concentrations and carrying capacity associated with woodheater emissions in Launceston, Tasmania. Atmos Environ 40(29):5543–5557

Luhar AK, Venkatram A, Lee SM (2006b) On relationships between urban and rural near-surface meteorology for diffusion applications. Atmos Environ 40(34):6541–6553

Malfroy H, Cope ME, Nelson PF (2005) An assessment of the contribution of coal-fired power station emissions to atmospheric particle concentrations in NSW a report prepared for: Delta Electricity, Eraring Energy and Macquarie Generation. Malfroy Environmental Strategies; CSIRO Energy Technology & CSIRO Atmospheric Research; Macquarie University Graduate

School of the Environment, Sydney, NSW. iv, 95 p. Available: http://www.cmar.csiro.au/e-print/open/cope_2004b.pdf

Martin G et al (2004) Evaluation of the atmospheric performance of HadGAM/GEM1. Hadley Centre Technical Note 54, Sep 2004. Available: http://www.metoffice.gov.uk/research/hadleycentre/pubs/HCTN/index.html

McGregor JL (1997) Semi-Lagrangian advection on a cubic gnomonic projection of the sphere. In: Lin C, Laprise R, Ritchie H (eds) Numerical methods in atmospheric and oceanic modelling: the André J. Robert Memorial (Companion volume to atmosphere-ocean). Canadian Meteorological and Oceanographic Society, Ottawa, Canada, pp 153–169

McGregor JL, Dix MR (2005) The conformal-cubic atmospheric model: progress and plans. In: Workshop on high resolution atmospheric simulations and cooperative output data analysis, Yokohama, Japan. JSPS International Meeting Series. JSPS and JAMSTEC, Yokohama, 2 p. Available: http:// www.es.jamstec.go.jp/esc/research/AtmOcn/hires2005/abstract/4-2_mcgregor.pdf

McGregor JL, Nguyen KC, Thatcher MJ (2007) Current CCAM modelling activities. AMOS 2007: 14th National Australian Meteorological and Oceanographic Society (AMOS) Conference in conjunction with Southern [Annular] Mode (SAM) Workshop: 5–8 February 2007, Adelaide, South Australia: climate water, and marine forecasting: challenges for the future [abstracts]., Adelaide, South Australia. Melbourne: AMOS. p 71. Available: http://www.cmar.csiro.au/e-print/open/cope_2005e.pdf

Physick WL, Noonan JA, McGregor JL, Hurley PJ, Abbs DJ, Manins PC (1994) LADM: a Lagrangian atmospheric dispersion model. CSIRO, Aspendale. CSIRO Division of Atmospheric Research technical paper; no. 24. 137 p. Available: http://www.cmar.csiro.au/e-print/open/physick_1994a.pdf

Physick WL, Cope ME, Lee SH, Hurley PJ (2006) Optimum exposure fields for epidemiology and health forecasting. 28th NATO/CCMS international technical meeting on air pollution modelling and its application: preprints: 15–19 May 2006, Leipzig, Germany. The Committee, Leipzig, pp 426–433

Physick WL, Cope ME, Lee SH, Hurley PJ (2007) An approach for estimating exposure to ambient concentrations. J Expo Sci Environ Epidemiol 17:76–83

Tory K, Cope ME, Hess GD, Lee SH, Puri K, Manins PC, Wong N (2004) The Australian air quality forecasting system. Part III: case study of a Melbourne 4-day photochemical smog event. J Appl Meteorol 43(5):680–695

Wain AG, Lee SH, Mills GA, Hess GD, Cope ME, Tindale N (2006) Meteorological overview and verification of HYSPLIT and AAQFS dust forecasts for the duststorm of 22–24 October 2002. Aust Meteorol Mag 55(1):35–46

Chapter 13
Coupling of Air Quality and Weather Forecasting: Progress and Plans at met.no

Viel Ødegaard, Leonor Tarrasón, and Jerzy Bartnicki

13.1 Introduction

Air quality modelling at met.no consists of three different systems, all coupled off-line to our numerical weather prediction (NWP) models. These are: (1) a nuclear emergency system, (2) an urban air quality (AQ) forecasting system and (3) a long-term air quality chemical transport model routinely used in Europe to determine transboundary pollution fluxes.

The first system, the "Severe Nuclear Accident Program" (SNAP) model was developed at met.no to allow emergency risk assessment (Saltbones et al. 1995, 1998). This is a Lagrangian particle model transporting gases, noble gases, particles of different size and density. The modeled processes are advection and diffusion by random walk, dry deposition with gravitational settling velocity parametrization for particles and wet deposition as function of size and precipitation for particles. The model is operated by forecasters and the Norwegian Radiation Protection Authority (NRPA) in case of nuclear accident. It runs on meteorological input from operational HIgh Resolution Limited Area Model (HIRLAM) (10 and 20 km horizontal resolution) and from ECMWF (Bartnicki et al. 2005).

The second system, the urban air quality information system runs operationally at met.no and consists of the chemical dispersion model AirQUIS developed at Norwegian Institute for Air Research (NILU) and the non-hydrostatic NWP model MM5 in 1 km horizontal resolution nested in HIRLAM (Berge et al. 2002). AirQUIS is an Eulerian gridpoint model with point source emissions, line source emissions and area source emissions. The prognostic components of the model are PM_{10}, $PM_{2.5}$ and nitrogen dioxide (NO_2). The system runs daily for 48 h forecasts for six Norwegian cities, with main focus in forecasting of urban air quality during winter season. The cities under study are located in low elevated areas surrounded by hills and mountains. Winter time inversions inhibit ventilation

V. Ødegaard (✉)
Norwegian Meteorological Institute (DNMI, met.no), Postboks 43, Blindern, N-0313 Oslo, Norway
e-mail: v.odegaard@met.no

A. Baklanov et al. (eds.), *Integrated Systems of Meso-Meteorological and Chemical Transport Models*, DOI 10.1007/978-3-642-13980-2_13,
© Springer-Verlag Berlin Heidelberg 2011

of pollution and thus, main exceedances of critical pollution levels occur in Norway during that time. The forecasts are distributed to end-users via newspapers, mobile network, and internet.

The third system is the ACTM supporting the modelling work under the Co-operative programme for monitoring and evaluation of the long-range transport of air pollutants in Europe (the EMEP programme). The model used, the EMEP Unified model, is developed at met.no for simulating atmospheric transport and deposition of acidifying and eutrophying compounds, aerosols as well as photo-oxidants over Europe. The model is a multi-layer Eulerian model and is now flexible with respect to the choice of horizontal grid projection, domain, and resolution. The model can thus, be run at local, regional, hemispheric, and global scales. Typically, the model simulates 1 year period of the transport and the current results of the regional model runs are available for the years 1980, 1985, 1990 and each year from 1995 to 2004. The EMEP Unified model in operational configuration is the regional version that uses presently HIRLAM PS (a dedicated polar-stereographic version of HIRLAM, which has been frozen for last 10 years) which runs on a polar stereographic grid with a 50 km resolution and covers Europe and the Atlantic Ocean. In vertical, the model has 20 sigma layers reaching up to 100 hPa. Approximately ten of these layers are below 2 km in order to obtain high resolution of the boundary layer. The polar stereographic projection is historically bound to the EMEP reporting grid for emissions over Europe. A detail description of the model can be found in Simpson et al. (2003) and validation results are available in Tarrasón (2003). The EMEP model is also run at regional scale in forecast mode with input from ECMWF Integrated Forecasting System (IFS). Initial results from these forecasts that have been running for 10 months are under evaluation and will be reported in due time. The hemispheric and global versions of the model are run based on ECMWF ERA data and ECMWF IFS archived data, respectively. A summary of the model performance in both hemispheric and global scales can be found in Jonson et al. (2006, 2007).

13.2 Off-Line Coupling of Meteorological and Chemical Transport Models

The Severe Nuclear Accident Program (SNAP) model runs with input from available operational HIRLAM. At present, the version runs on 20 km horizontal resolution and provides a sufficiently large domain for SNAP. The most important meteorological input is 3D wind, precipitation and temperature fields. The time resolution for meteorological input is 3-h in operational applications, and 1-h – historical simulations.

The AirQUIS model runs with input from MM5 (Berge et al. 2002) in 1-h time resolution. The transferred 2D surface parameters are precipitation, total cloud cover, mixing height and surface temperature. 2D parameters from lowest model level are the air temperature, dew point, relative humidity, and vertical temperature

Table 13.1 Variables transferred from HIRLAM to EMEP (from Simpson et al. 2003)

Parameter	Unit	Description	Main purpose
3D fields for 20 σ levels			
u,v	m/s	Wind velocity components	Horizontal advection
Q	kg/kg	Specific humidity	Chemical reactions, dry deposition
$\dot{\sigma}$	s^{-1}	Vertical wind in σ coordinate	Vertical advection
θ	K	Potential temperature	Chemical reactions, eddy diffusion
CL	%	Cloud cover	Wet removal. photolysis
PR	med mer	Precipitation	Wet and dry deposition
2D fields for surface			
P_S	hPa	Surface pressure	Surface air density
T^2	K	Temperature at 2 m height	Dry deposition, stability
H	Wm^{-2}	Surface flux of sensible heat	Dry deposition, stability
τ	Mm^{-2}	Surface stress	Dry deposition, stability
LH	Wm^{-2}	Surface flux of latent heat	Dry deposition

gradient. The horizontal wind is the only 3D parameter used. A horizontal interpolation from polar stereographic grid to Universal Transverse Mercator (UTM) is taking place. At present no vertical interpolation is done. The vertical levels in AirQUIS are defined to be identical to the levels in MM5. A meteorological pre-processor calculates dispersion parameters based on the Monin–Obukhov similarity theory (MOST). The constant parameters topography and surface roughness are taken from the meteorological model.

The meteorological input required in the Unified EMEP model are the 3D horizontal and vertical wind fields, specific humidity, potential temperature cloud cover, and precipitation. The transferred surface 2D fields for use in the chemical transport model are: surface pressure, 2 m temperature, surface flux of momentum, sensible and latent heat, and surface stress. All variables are given in 3-h interval. Table 13.1 lists the variables and their main purposes in the EMEP model. Inside the model different boundary layer parameters like the stability, eddy diffusion, and mixing height are calculated based on MOST.

13.3 Evaluation of Urban Air Quality Forecasts

Regular forecasts are produced with MM5/AirQUIS. The meteorological and air quality forecasts are evaluated against observations and reported on a yearly basis. Summary statistics and case studies are produced. In Fig. 13.1 forecasts for the AQ station Alnabru are compared to AQ observations at the same station and to meteorological observations at the closest station Valle Hovin. The missing peak in the NO_2 forecasts is not caused by errors in the meteorological forecasts. However, the AQ monitoring station is not located together with the meteorological station. Neither does the meteorological station measure all the parameters that used by the AQ model. Observations of inversion layer are limited to measurements of air temperature at 2 and 25 m in the presented case.

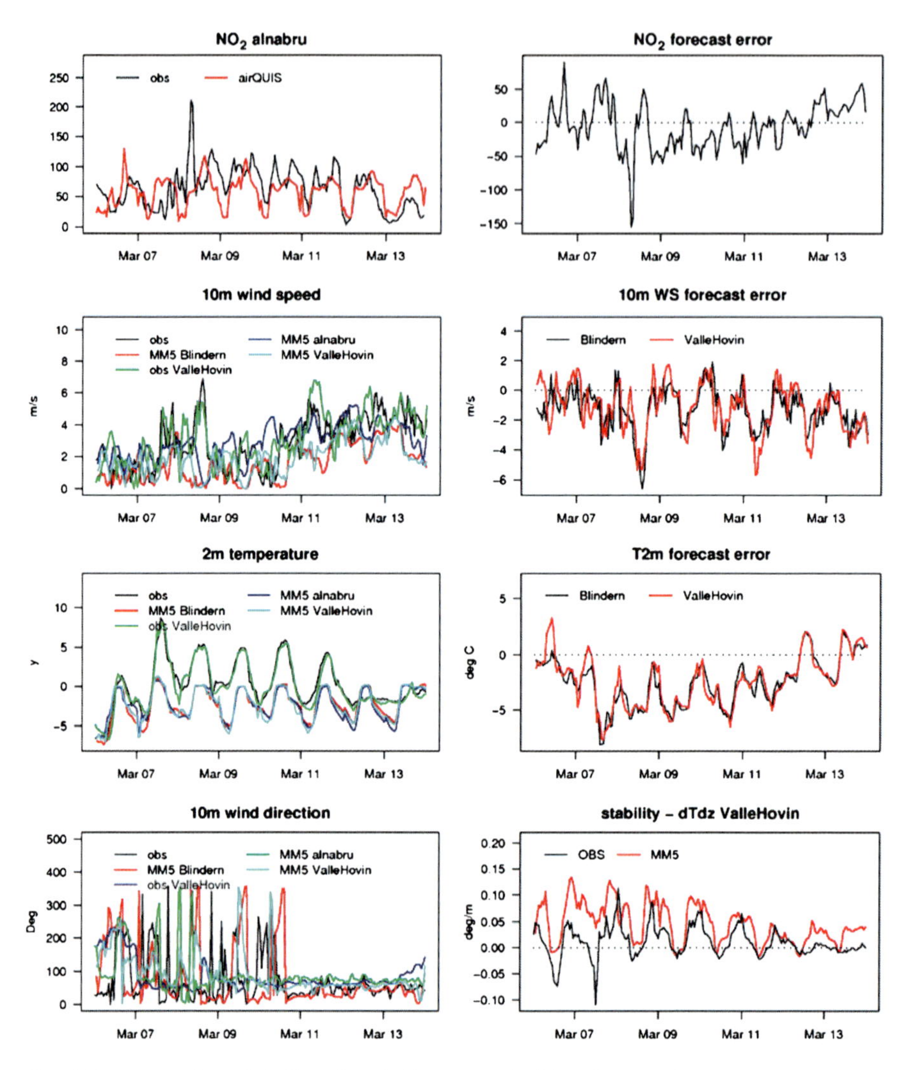

Fig. 13.1 Observations (*black*) and model forecasts of NO_2, wind speed, 2 m temperature and wind direction (*left*), and forecast error for NO_2, wind speed and 2 m temperature (*right*) at observation site Valle Hovin. Forecasts (*dotted*) and observations (*solid*) of vertical temperature gradient 2–25 m (*bottom right*) (from Ødegaard et al. 2004)

Experiments have been performed to address the error made by the pre-processor in AirQUIS. Figure 13.2 shows the resulting NO_2 forecasts from AirQUIS using dispersion parameters calculated by the pre-processor compared to AirQUIS using dispersion parameters calculated by the meteorological model MM5.

Both in the pre-processor and in MM5 the dispersion parameters are calculated using MOST. The difference is, therefore, due to a time step update of the parameters going into the parametrization scheme rather than an hourly update in the pre-processor. The figure shows that this difference has very small impact on the results.

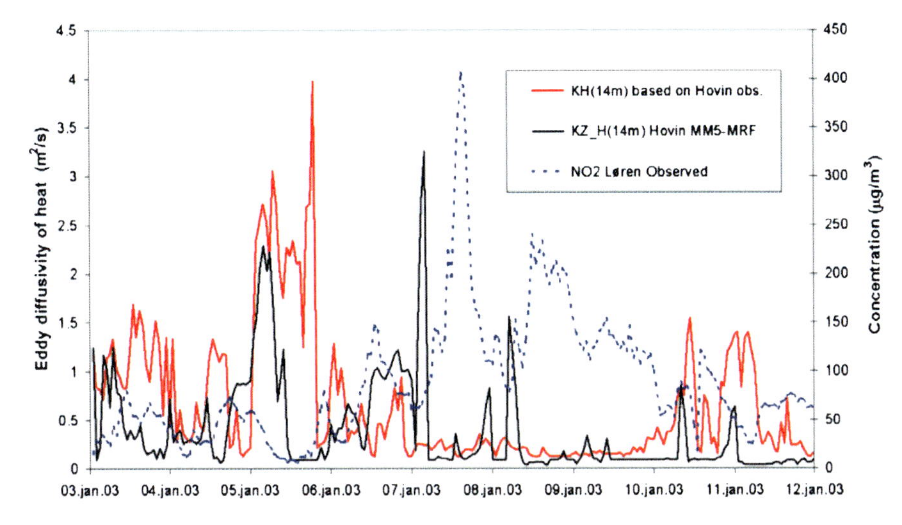

Fig. 13.2 NO_2 forecasts from AirQUIS using surface fluxes from pre-processor (*black*) and from MM5 (*blue*) compared to observations (*red*) (from Slørdal and Ødegaard, 2005)

13.4 Issues to Consider for NWP Models Providing Data for Air Pollution Models

All air pollution modelling systems at met.no are presently developing to include other types of meteorological drivers at finer resolution. For instance, a small scale SNAP version for simulating local effects in the range of 30–50 km is planned for 2007. Further a coupling to UK Met Office Unified model (UM) or small scale HIRLAM model will be made for even higher resolution SNAP. Also a full coupling of SNAP to regional HIRLAM model is in the line, i.e. SNAP as subroutine in HIRLAM code. During the winter season 2006–2007 the UM model was introduced to replace MM5, and a new interface to AirQUIS is built. Also the EMEP model is developing to use different main sets of meteorological data, ECMWF, HIRLAM and the non-hydrostatic models Weather Research and Forecasting Model (WRF), UM, and ALADIN.

For all these applications, special attention should be given to the interfaces between dynamical and chemical processes. While it is recognised that an on-line coupling of these processes will be ultimately necessary, there is still a series of processes that need special attention also under off-line applications, as named below.

13.4.1 Surface Classes

For consistent calculation of boundary layer parameters in off-line coupling the land-use classes in the meteorological model should ideally match the land-use classes presented in EMEP (Table 13.2). In the coupling with HIRLAM, where only

Table 13.2 Surface types used by the EMEP model

Surface/vegetation type	H (m)	Albedo (%)
Temperate/boreal coniferous forests	>20	12
Temperate/boreal deciduous forests	>20	16
Mediterranean needle-leaf forests	15	12
Mediterranean broadleaf forests	15	16
Temperate crops	1	20
Root crops	1	20
Mediterranean crops	2	20
Semi-natural moorland	0.5	14
Grassland	0.5	20
Mediterranean scrubs	3	20
Wetlands	0.5	14
Tundra	0.5	15
Desert	0	25
Water	0	8
Ice	0	70
Urban	10	18

five land-use classes are presented, and no parameters to distinguish needle leaf from broad leaf forest, is thus not fully consistent.

13.4.2 Physical Parametrizations

ACTM models can make use of atmospheric parameters that are output from some parametrization schemes. Boundary layer parameters have to be calculated inside ACTM models or in pre-processing if they are not available from the meteorological model. Entrainment and detrainment rates in cumulus clouds could be provided if sufficiently sophisticated cumulus parametrization is used in the meteorological model.

References

Bartnicki J, Salbu B, Saltbones J, Foss A, Lind OC (2005) Atmospheric transport and deposition of radioactive particles from potential accidents at Kola nuclear power plant. Re-analysis of worst case scenarios. Norwegian Meteorological Institute, Oslo, Norway. Met.no research report No. 10/2005. ISSN-1503-8025

Berge E, Walker SE, Sorteberg A, Lenkopane MS, Jablonska HTB, Køltzow MØ (2002) A real-time operational forecast model for meteorology and air quality during peak air pollution episodes in Oslo, Norway. Water Air Soil Pollut Focus 2:745–757

Jonson JE, Wind P, Gauss M, Tsyro S, Søvde AO, Klein H, Isaksen ISA, Tarrasón L (2006) First results from the hemispheric EMEP model and comparison with the global Oslo CTM2 model, EMEP/MSC-W note 2/2006. Norwegian Meteorological Institute, Oslo, Norway

Jonson JE, Tarrason L, Wind P, Gauss M, Valiyaveetil S, Tsyro S, Klein H, Isaksen ISA, Benedictow A (2007) First evaluation of the global EMEP model and comparison with the global Oslo CTM2 model, EMEP/MSC-W Technical Report 2/2007. Norwegian Meteorological Institute, Oslo, Norway

Saltbones J, Foss A, Bartnicki J (1995) Severe nuclear accident program (SNAP) – a real time dispersion model. In: Gryning SE, Schiermeier FA (eds) Proceedings of the 21st NATO/CCMS Meeting on air pollution modelling and its application, November 1995, Baltimore, USA, pp 333–340

Saltbones J, Foss A, Bartnicki J (1998) Norwegian Meteorological Institute's Real-Time Dispersion Model SNAP (Severe Nuclear Accident Program). Runs for ETEX and ATMES II Experiments with Different Meteorological Input. Atmos Environ 32(24):4277–4283

Simpson D, Fagerli H, Jonson JE, Tsyro S, Wind P, Tuovinen J-P (2003) Transboundary acidification, eutrophication and ground level ozone in Europe, Part I. Model Description. EMEP Status Report, 1/2003. Norwegian Meteorological Institute, Oslo, Norway

Slørdal H, Ødegaard V (2005) Dispersion conditions in the stable boundary layer as described by the MM5-model – a case study of a pollution episode in Oslo, Norway. Proceedings of 5th international conference on urban air quality, 29–31 Mar 2005, Valencia, Spain

Tarrasón L (eds) (2003) Transboundary acidification, eutrophication and ground level ozone in Europe, Part II. Model Validation. EMEP Status Report, 1/2003. Norwegian Meteorological Institute, Oslo, Norway

Ødegaard V, Gjerstad KI, Bjergene N, Jablonska HTB, Walker SE (2004) Evaluation of a forecast model for meteorology and air quality winter 2003/2004 (In Norwegian). Norwegian Meteorological Institute. Oslo, Norway. Met.no research report No. 12/2004. ISSN-1503-8025

Chapter 14
A Note on Using the Non-hydrostatic Model AROME as a Driver for the MATCH Model

Lennart Robertson and Valentin Foltescu

14.1 Introduction

Mass conservation is one of the desired properties of transport schemes (Williamson 1992). The design of the transport scheme as such is one of the parts to achieve this, but is necessarily not enough as inconsistent meteorological data may not fulfill mass conservation. The MATCH model is an off-line model with several options on output grids and flexible on input data on various resolutions. The vertical coordinates and resolution are however, adopted from the input meteorological data and restricted to hybrid sigma-pressure coordinates. The transport scheme is a modified flux oriented Bott scheme (Robertson et al. 1998). Another feature of the model is that the vertical winds are calculated internally of two major reasons: (1) the relative vertical wind is needed (normally not available in driving data), (2) internal interpolation of meteorological data (in time and space) demands recalculation of the relative vertical winds (Robertson et al. 1998). The vertical wind calculation is very sensitive to errors in mass divergence that do not correspond to pressure tendencies. A procedure for initialization is therefore, implemented based on the methodology proposed by Heiman and Keeling (1989), where the horizontal winds are iteratively corrected until the vertically integrated mass divergence corresponds to the pressure tendency. The methodology has an inherent assumption of hydrostatic balance. The inconsistency in input data may arise from e.g. spectral numerical weather prediction (NWP) data interpolated to a grid, or induced by internal interpolations in time and space. There is always an interpolation associated with data from spectral NWP models provided on a grid-mesh, coming from the need of staggering horizontal wind components (to Arakawa C grid) demanded by the flux oriented transport scheme. An additional source of inconsistency is truncation by compression of data in e.g. GRIdded

L. Robertson (✉)
Swedish Meteorological and Hydrological Institute (SMHI), SE-601 76 Norrköping, Sweden
e-mail: lennart.robertson@smhi.se

A. Baklanov et al. (eds.), *Integrated Systems of Meso-Meteorological and Chemical Transport Models*, DOI 10.1007/978-3-642-13980-2_14,
© Springer-Verlag Berlin Heidelberg 2011

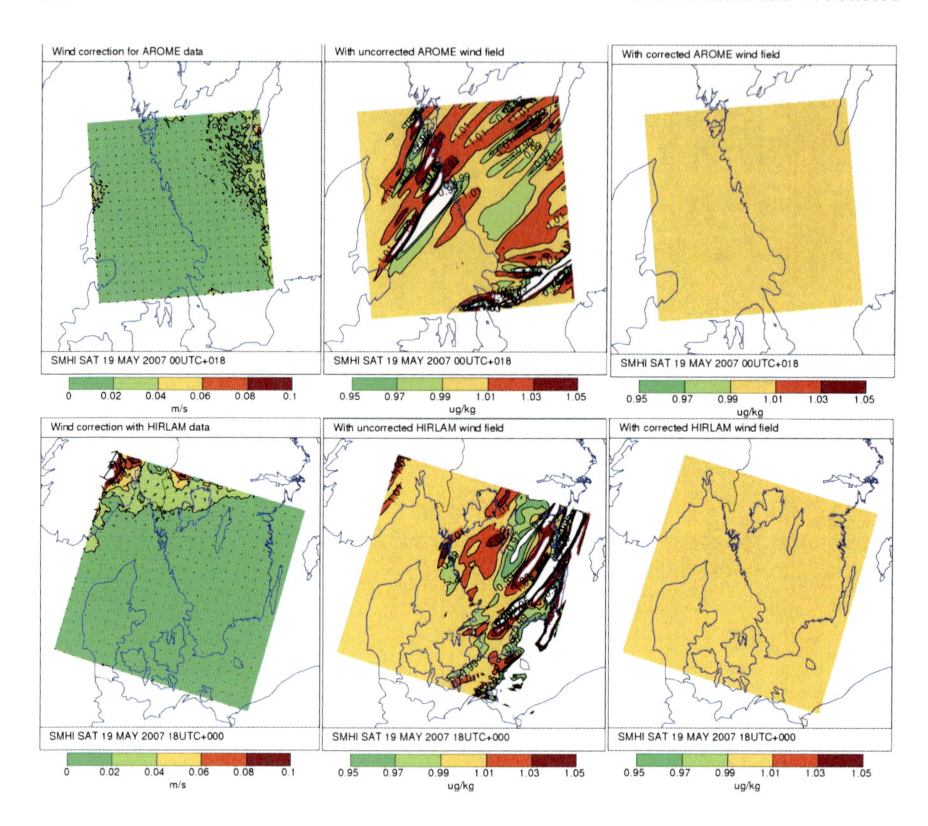

Fig. 14.1 (*Top panel*): magnitudes of changes of the horizontal wind fields during initializations for AROME (*left*) and HIRLAM (*right*) data; (*Middle panel*): mass conservation test then using non-hydrostatic AROME model data (2.5 km resolution) over a sub-area covering the Swedish west coast - simulations over 18 h without (*left*) and with (*right*) initialization of horizontal winds (the extreme values of the errors in the left panel ranges between −40 to + 100%); (*Bottom panel*): mass conservation test then using the hydrostatic HIRLAM model data (5 km resolution) over a sub-area covering southern Scandinavia – simulations over 18 h without (*left*) and with (*right*) initialization of horizontal winds (the extreme values of the errors in the left panel ranges between −30 to + 90% and probably associated with a frontal zone passage)

Binary (GRIB) format. From this perspective non-hydrostatic NWP data will appear as "inconsistent" data.

14.2 Mass Conservation Test Run

It is rather straight forward to check the ability of mass conservation of the transport scheme, by initializing the model domain with a constant mixing ratio internally and on the boundaries, and during model run the mixing ratio should stay constant.

The test setup works for schemes on flux-form but does not apply to semi-Lagrangian schemes (interpolation of equal numbers) or schemes on advective form (difference of equal numbers).

Figure 14.1 (middle and bottom panels) show test runs over 18 h for non-hydrostatic driving data (AROME 2.5 km) and hydrostatic data (HIRLAM 5 km). The mass errors with no initialization ranges in both cases from −30 to + 100% (which means a rather fast deterioration of the mass conservation).

In Fig. 14.1 (top panel) the impact from the initialization on the horizontal winds is illustrated. The modifications are generally less than 0.1 m/s and over large areas less than 0.02 m/s, both for hydrostatic and non-hydrostatic data. The corrections are thus small and could not be expected to violate the general flow pattern. Note that the corrections seem to be very small in relation to the corresponding errors in Fig. 14.1 (middle and bottom panels), which is a bit misleading as the latter results from integrated errors.

14.3 Conclusions

We have illustrated the need for initialization of non-hydrostatic as well as hydrostatic driving meteorological data for off-line atmospheric chemistry and transport models. The impact on the wind field from initialization is of the same magnitudes for both non-hydrostatic and hydrostatic data (i.e. less than a few dm/s), that indicates that no specific problem concerning initialization for mass conservation of non-hydrostatic data.

References

Heiman M, Keeling DD (1989) A three dimensional model of atmospheric CO_2 transport based on observed winds. 2. Model description and simulated tracer experiments. In: Peterson DH (ed) Aspects of climate variability in the pacific and western Americas. Am Geophys Union, pp 237–275

Robertson L, Langner J, Engardt M (1998) An Eulerian limited-area atmospheric transport model. J Appl Meteorol 38:190–210

Williamson DL (1992) Review of numerical approaches for modelling of global transport. In: van Dop H, Kallos G (eds) Proceedings of the 19th NATO/CCMS ITM on air pollution modelling and its application, Sept 29–Oct 4, pp 377–398

Chapter 15
Aerosol Species in the Air Quality Forecasting System of FMI: Possibilities for Coupling with NWP Models

Mikhail Sofiev and SILAM Team

15.1 Introduction

The regional air quality (AQ) forecasting system of FMI has been set up in 2005 and opened for public access via internet in 2006 (http://silam.fmi.fi). A primary goal of the system is to evaluate and forecast the air pollution over the Finnish territory. Since Finland is a receptor of practically all main pollutants, the area of the simulations is necessarily covered the whole of Europe with a compromising resolution of 30 km. A nested domain was introduced in 2007 and it covers northern Europe with a resolution of 10 km.

The pollutants of primary concern are: particulate matter, first of all, fine particles, nitrogen oxides, and allergenic pollutants, such as birch pollen. Since in most cases the problems with NOx species are confined to Helsinki, the regional forecasting system originally did not include nitrogen chemistry limiting the chemical simulations with the sulphur oxides as an indicator of anthropogenic plumes. Currently, the new extension of the AQ forecasts is on the way to include operational simulations for all main anthropogenic pollutants including SOx, NOx, NHx, O_3, and VOCs. Corresponding setup has been created and is being tested in trial operational simulations.

15.2 Materials and Methods

15.2.1 The Forecasting System

Since February of 2006, the regional forecasting system of FMI covers three major types of sources (Fig. 15.1): anthropogenic emission of sulphur oxides and primary particulate matter PM 2.5 and PM 10, biological sources of birch pollen and

M. Sofiev (✉)
Finnish Meteorological Institute (FMI), P.O. Box 503, 00101 Helsinki, Finland
e-mail: mikhail.sofiev@fmi.fi

A. Baklanov et al. (eds.), *Integrated Systems of Meso-Meteorological and Chemical Transport Models*, DOI 10.1007/978-3-642-13980-2_15,
© Springer-Verlag Berlin Heidelberg 2011

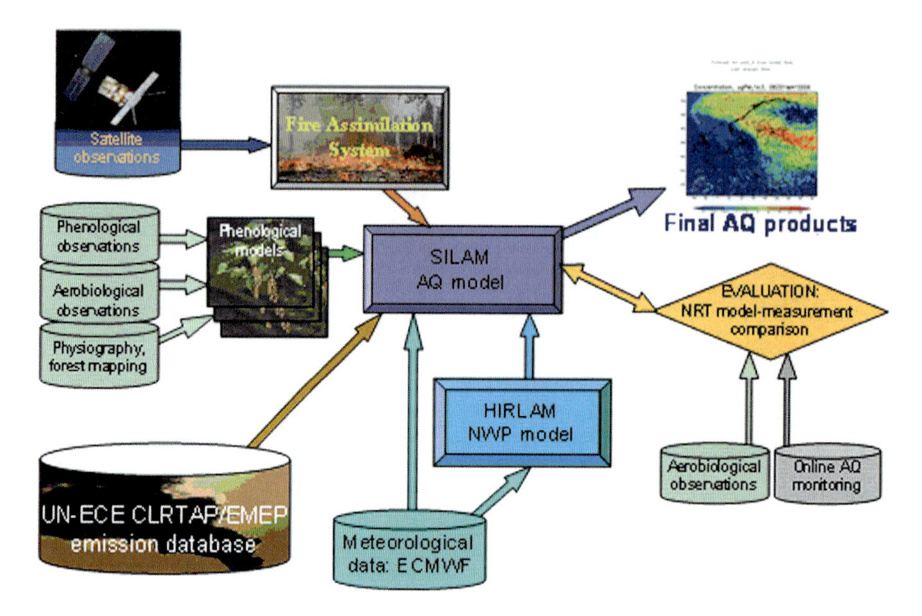

Fig. 15.1 A structure and main items of the regional air quality forecasting system of FMI

satellite-retrieved real-time information about the wild-land fires (based on hot-spots counts from MODIS instrument onboard NASA Aqua and Terra spacecrafts).

15.2.2 The SILAM Model

SILAM is a dual-core modelling system with Lagrangian dispersion core based on an iterative advection algorithm of Eerola (1990) and a Monte Carlo random-walk diffusion representation (Sofiev et al. 2006b), and an Eulerian dynamic core that applies the advection routine of Galperin (2000) with vertical diffusion scheme based on extended resistive analogy of Sofiev (2002) and parametrization of vertical diffusivity after Genikhovich et al. (2007). The system can directly utilize the meteorological data from the HIRLAM and ECMWF numerical weather prediction (NWP) models, as well as their archives. A typical time step accepted for the operational forecasts and most of hindcast studies is 3 h.

The operational forecasts included the following pollutants and source categories:

- Sulphur oxides originated from anthropogenic and volcanic sources
- Primary particulate matter originated from anthropogenic sources (both fine particulate matter with diameter below 2.5 μm, $PM_{2.5}$, and the coarse fraction with diameter from 2.5 to 10 μm $PM_{2.5-10}$)
- Primary $PM_{2.5}$ originated from biomass burning
- Birch pollen originated from birch forests during the flowering season

Input emission data for the anthropogenic pollutants and Etna volcano are based on the database of the European Monitoring and Evaluation Programme (EMEP, http://www.emep.int).

The near-real-time information on active biomass burning is extracted from the observations of the MODIS instrument onboard the NASA Aqua and Terra satellites (http://modis.gsfc.nasa.gov) with a spatial resolution of 1 × 1 km. The emission fluxes of $PM_{2.5}$ are defined from two algorithms. One is based on scaling the absolute temperatures of grid cells labelled as burning using their temperature anomalies (the temperature anomaly is defined as the difference of the observed and the long-term average temperatures) following the procedure described in Saarikoski et al. (2007). The new method (currently under evaluation) is based on scaling the Fire Radiative Power following the methodologies of Ichhoku and Kaufman (2005) and Kaufman et al. (2003).

Pollen model follows the approach suggested by (Sofiev et al. 2006a). The start day of the release is additionally adjusted to the conditions of the specific year using the near-real-time pollen observations of the European Aeroallergen Network (EAN, http://www.univie.ac.at/ean/). Specifics of the pollen grains as an atmospheric pollutant are taken into account in the parametrizations of the dry and wet deposition.

15.3 Evaluation of the Forecasting System

During pre-operational SILAM evaluation its results were compared with the data of the European Tracer Experiment (Sofiev et al. 2006b), Chernobyl and Algeciras accidental releases. For AQ related species, the system output was compared with the European AQ networks for the period of 2000–2002 and several campaigns and for some extreme episodes, such as the case of wild-land fires during spring and summer 2006.

An example of the long-term comparison for sulphates is shown in Figs. 15.2 and 15.3, which are based on the year-long juxtaposition of the SILAM results and EMEP observational network. Comparing the results with e.g. EMEP simulations, one can conclude that SILAM accuracy is typical for such type of the simulations and the main features of distribution of anthropogenic sulphates are captured by the model. There are several problematic regions, however, such as Danish Straits and southern Baltic where the concentrations are substantially lower than the observed levels. Comparison of observed and modelled SO_2 concentrations and wet deposition (not shown) suggest a general deficit of SOx in this part of the model domain. One of possible reasons could be uncertain ship emission in the area.

Comparison for total particulate matter is more problematic. Cases with dominant fire-induced pollution are reproduced comparatively well (Fig. 15.4) while the contributions of dust and secondary organic aerosol are currently missing from the system. The resulting problems are illustrated by the Fig. 15.5, which represents a full-year time series for a continental EMEP station from Germany. It is seen that

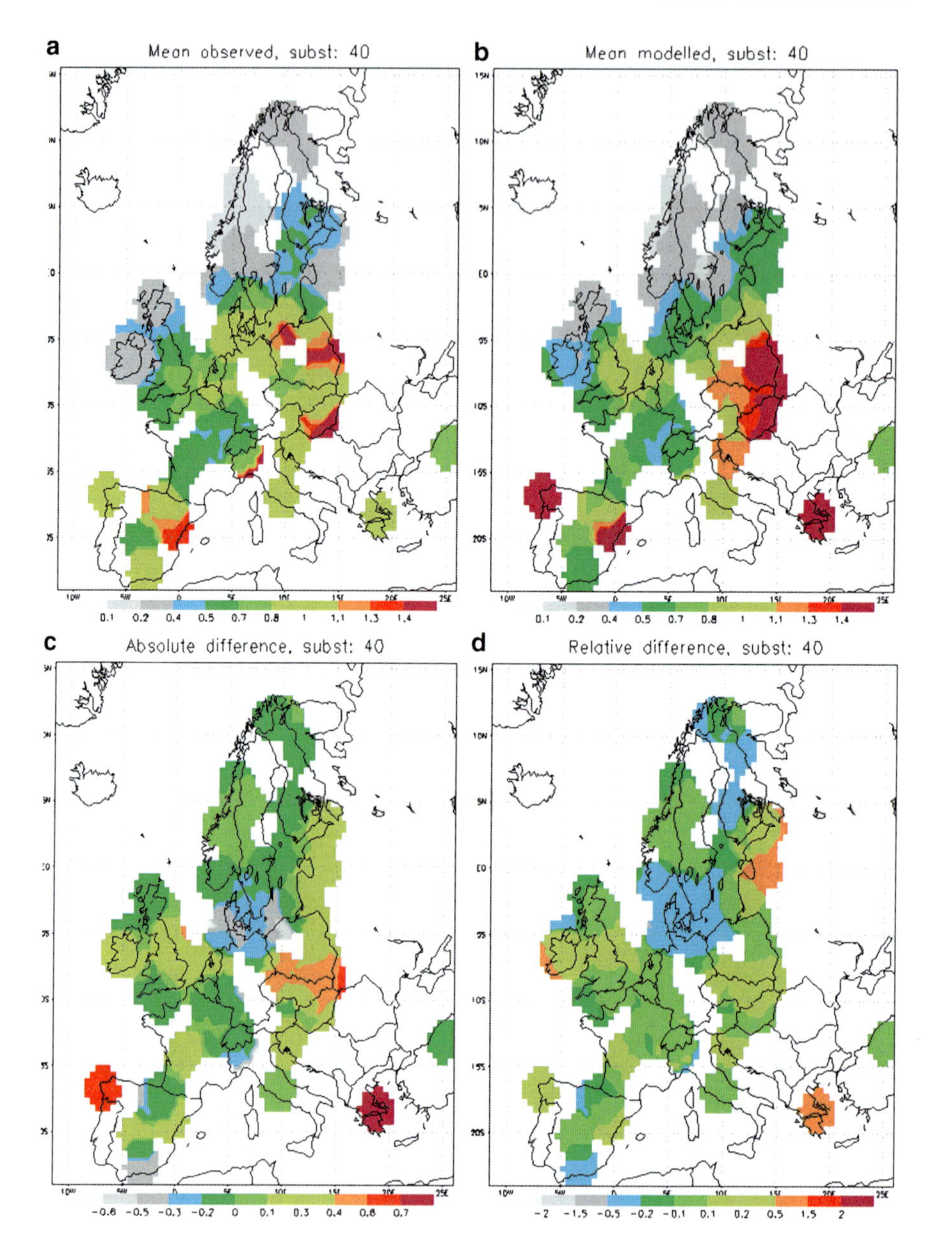

Fig. 15.2 Mean observed (**a**) and calculated (**b**) concentrations for sulphates in 2000, and absolute (**c**) and relative (**d**) differences. Unit: $\mu g\ S\ m^{-3}$

for wet and cold seasons the model results for PM_{10} are very good, as well as the sulphate concentration throughout the year. However, during the period of strong agricultural activity the system misses more than 50% of PM_{10}.

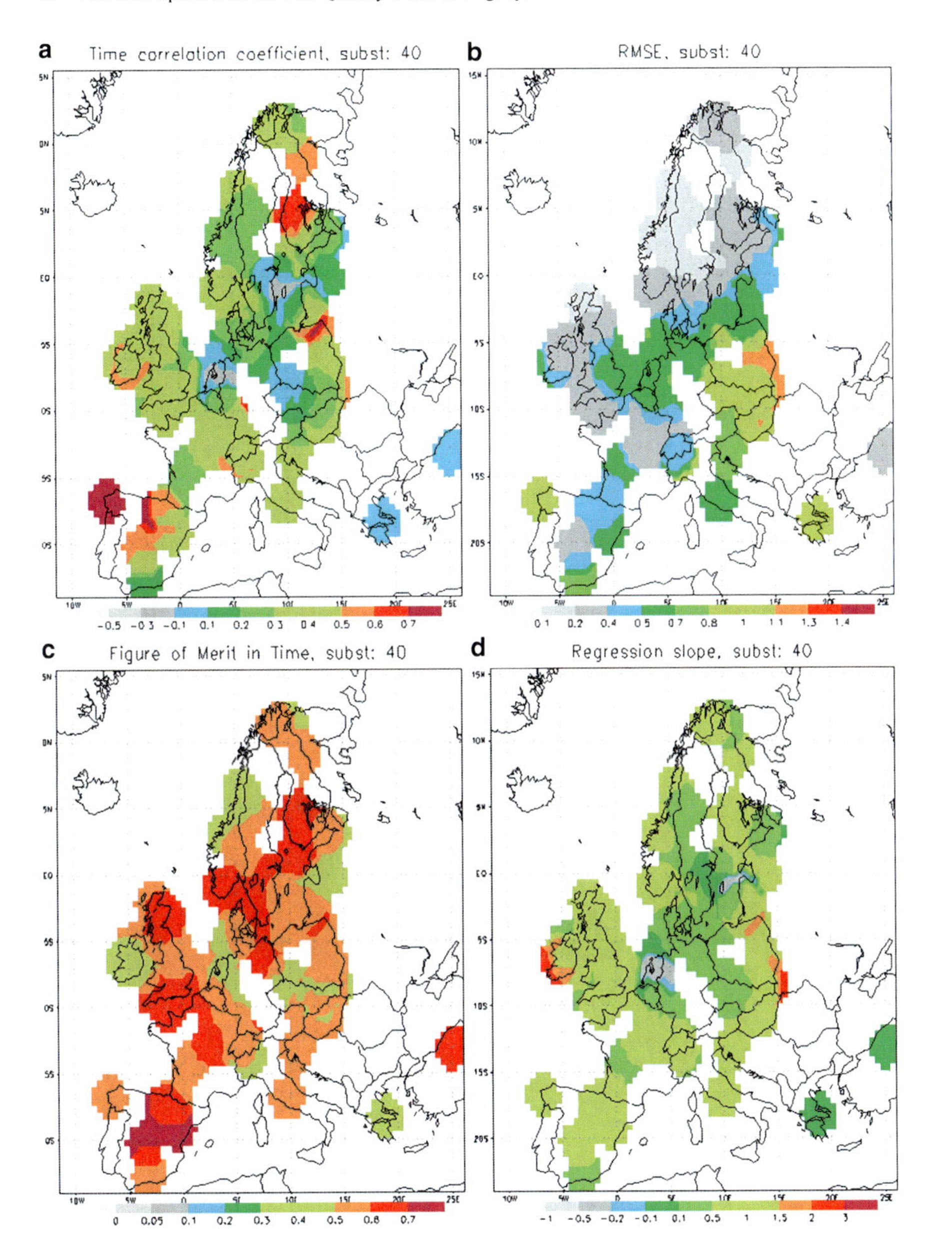

Fig. 15.3 Correlation coefficient over time (**a**), RMSE (**b**), Figure of Merit in Time (**c**) and regression slope (**d**) for sulphates in 2000. Unit: µg S m^{-3} for RMSE, relative for the other measures

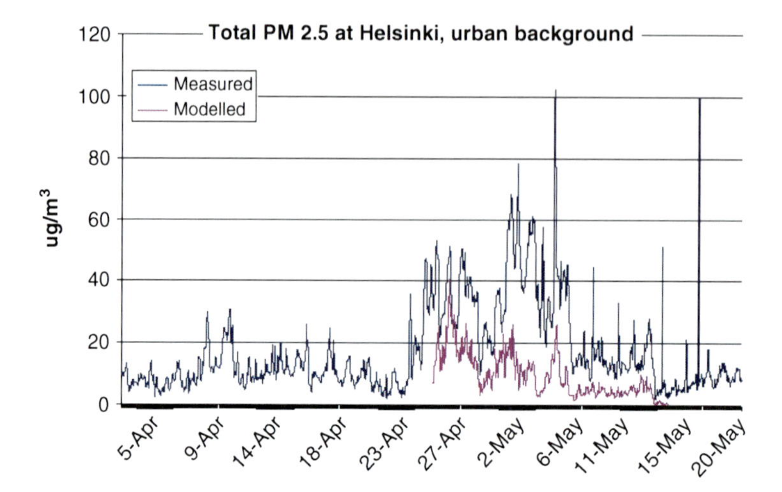

Fig. 15.4 Comparison of predicted $PM_{2.5}$ from fires and anthropogenic sources with total-$PM_{2.5}$ observed at Kumpula station in April–May 2006 (urban background in Helsinki). Peaks are attributed to fire smoke

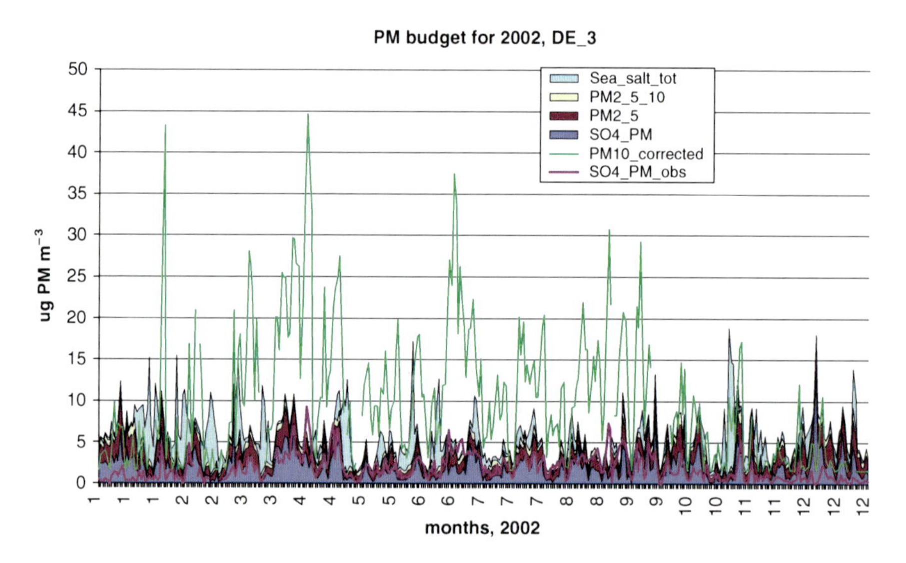

Fig. 15.5 Comparison of observed PM_{10} and total modelled PM_{10} composition for 2002. Unit: μg PM m^{-3}

15.4 Feedbacks with the Off-Line Coupled Modelling Systems

The above forecasting system, upon completion of tests of the new chemistry scheme, will be able to simulate up to 80–90% of fine-size atmospheric aerosols. Such coverage would make it feasible to consider the feedbacks to meteorological simulations. Consideration of such feedbacks using the existing modelling systems, which are usually independent and can only be interfaced with each other, is feasible but challenging task.

Technically, the transfer of the information from the ACTM to meteorological model is straightforward. For the operational forecasts, the next cycle meteorological calculations can use the previous cycle chemical composition results. For the case of nested simulations, both re-analyses and forecasts, the next-step nested meteorological computations can utilise the previous-step composition data.

The main challenge, however, comes from the other side. It is not enough to transfer the data from ACTM to NWP model, it is necessary to create the corresponding structures and algorithms inside the NWP model itself, so that it is capable of making use of this information. Most of classical NWP models, including HIRLAM, have far-reaching parametrizations of most of radiation transfer and cloud microphysics processes. These parametrizations do not account for any actual information on aerosol content or gaseous composition of the atmosphere. To the opposite, they are based on "average" numbers and tuned to ensure the best score of the meteorological forecasts. In such a situation, replacement of the well-tuned parametrizations deeply embedded into the model physics with something new will be quite laborious and inevitably worsen the quality scores, at least, at the beginning.

An alternative approach would be to utilise the models where some elements of external forcing is foreseen. Development of such systems can be easier.

A limitation of off-line feedback mechanisms is an implicit assumption that a single iteration NWP-ACTM-NWP is sufficient to reflect the bulk of the impact of chemical composition onto meteorology. This assumption is fulfilled in almost all cases but larger number of iterations might be needed in case of very strong deviation of the atmospheric composition from the default values assumed in the NWP model. Importance of this limitation and reasonable number of iterations needed for e.g. a dust storm simulations need investigation.

15.5 Conclusion

The current setup of the FMI chemical weather forecasting system follows the standard approach of the off-line coupling of meteorological and chemical transport models. It creates the possibility of utilising different meteorological drivers and to perform ensemble-type simulations. The forecasting system covers most types of the atmospheric aerosols, except for dust and secondary organic particles and shows good results in comparison against observations, both long-term and dedicated campaigns. Full-chemistry model version with nitrates and ammonium is at the final evaluation stage. Possibilities for the feedback from chemical composition to meteorology using the off-line models exists and comparatively straightforward from technical point of view. The obstacle, however, exits in the formulations of most of existing NWP models where the influence of the atmospheric composition is parameterised and hard-coded into the corresponding schemes, thus making it difficult to assimilate the chemical and aerosol data coming from the air quality models.

Acknowledgements The study was performed within the scope of EU-GEMS project (Global Earth-System Monitoring using satellite and in-situ data) and the ESA PROtocol MOniToring for the GMES Service Element: Atmosphere (PROMOTE) project (http://www.gse-promote.org). Pollen simulations are based on outcome of the Finnish Academy POLLEN project.

References

Eerola K (1990) Experimentation with a three-dimensional trajectory model, vol 15. FMI Meteorological, Helsinki, 33 p

Galperin MV (2000) The approaches to correct computation of airborne pollution advection, vol XVII, Problems of ecological monitoring and ecosystem modelling. Gidrometeoizdat, St. Petersburg, pp 54–68

Genikhovich E, Sofiev M, Gracheva I (2007) Interactions of meteorological and dispersion models at different scales. In: Borrego C, Norman A-L (eds) Air pollution modelling and its applications, vol XVII. Springer, New York, pp 158–166, ISBN-10: 0-387-28255-6

Ichhoku C, Kaufman YJ (2005) A method to derive smoke emission rates from MODIS fire radiative energy measurements. IEEE Trans Geosci Remote Sens 43(11):1–12

Kaufman YJ, Ichoku C, Giglio L, Korontzi S, Chu DA, Hao WM, Li R-R, Justice CO (2003) Fires and smoke observed from the earth observing system MODIS instrument – products, validation, and operational use. Int J Remote Sens 24(8):1765–1781

Saarikoski S, Sillanpää M, Sofiev M, Timonen H, Saarnio K, Teinilä K, Karppinen A, Kukkonen J, Hillamo R (2007) Chemical composition of aerosols during a major biomass burning episode over northern Europe in spring 2006: experimental and modelling assessments. Atmos Environ 41:3577–3589

Sofiev M (2002) Extended resistance analogy for construction of the vertical diffusion scheme for dispersion models. J Geophys Research – Atmosphere 107, D12, doi:10.1029/2001JD001233

Sofiev M, Siljamo P, Ranta H, Rantio-Lehtimäki A (2006a) Towards numerical forecasting of long-range air transport of birch pollen: theoretical considerations and a feasibility study. Int J Biometeorol. doi:10 1007/s00484-006-0027-x

Sofiev M, Siljamo P, Valkama I, Ilvonen M, Kukkonen J (2006b) A dispersion modelling system SILAM and its evaluation against ETEX data. Atmos Environ. doi:10.1016/j.atmosenv.2005.09.069

Chapter 16
Overview of DMI ACT-NWP Modelling Systems

Alexander Baklanov, Alexander Mahura, Ulrik Korsholm, Roman Nuterman, Jens Havskov Sørensen, and Bjarne Amstrup

16.1 Introduction

The model development strategy at the Danish Meteorological Institute (DMI) includes both off-line and on-line integrated models. Several Atmospheric Chemistry Transport (ACT) off-line modelling systems had been developed at DMI: the Multi trajectOry Ordinary differential Numerical (MOON), the Chemistry Aerosol Cloud (CAC), and the Danish EmeRgency MAnagement (DERMA), the on-line Environment – HIgh Resolution Limited Area (Enviro-HIRLAM), Urban Scale High Resolution and Microscale Model for Urban Environment (M2UE). All the off-line models can be integrated with the Numerical Weather Prediction (NWP) models such as DMI-HIRLAM model or other meteorological models, or utilize NWP output data as meteo-drivers. In Fig. 16.1a these deterministic systems are presented:

- The MOON is a Lagrangian chemistry model used to simulate the air pollution both in the boundary layer and free troposphere, and for stratospheric modelling (Gross et al. 2005; Madsen 2006)
- The CAC (Gross and Baklanov 2004) is a further development of the MOON, with main difference between the two mentioned that the CAC model includes aerosol formation and dynamics and more sofisticated chemistry
- The Enviro-HIRLAM is an on-line (Korsholm et al. 2008); it is also used for pollen modelling (Mahura et al. 2006; Rasmussen et al. 2006)
- The M2UE is a Computational Fluid Dynamics (CFD) 3D obstacle-resolved air flow model (Nuterman et al. 2008)
- The Danish Emergency Response Model of the Atmosphere (DERMA) (Sørensen 1998; Sørensen et al. 2007) used for evaluation of radioactive, chemical, and biological accidental releases into the atmosphere

A. Baklanov (✉)
Danish Meteorological Institute (DMI), Lyngbyvej 100, DK-2100 Copenhagen, Denmark
e-mail: alb@dmi.dk

A. Baklanov et al. (eds.), *Integrated Systems of Meso-Meteorological and Chemical Transport Models*, DOI 10.1007/978-3-642-13980-2_16,
© Springer-Verlag Berlin Heidelberg 2011

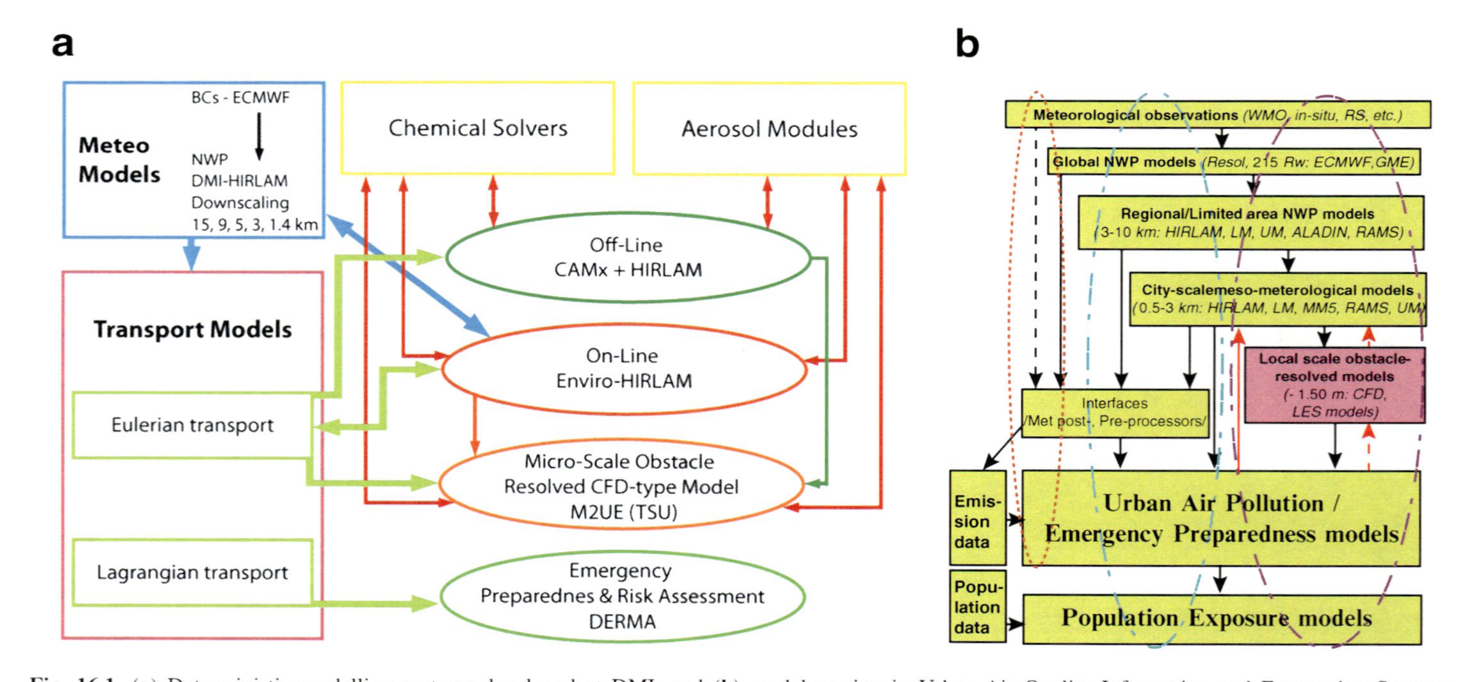

Fig. 16.1 (**a**) Deterministic modelling systems developed at DMI; and (**b**) model nesting in Urban Air Quality Information and Forecasting Systems (UAQIFS) (. . . common regulatory models; -.-.- FUMAPEX multi-scale systems; -..- new suggested down-scaling with obstacle-resolved models)

These modelling systems are developed in order to fulfil several duties at DMI, e.g. (1) smog and ozone preparedness, (2) nuclear, veterinary, and chemical emergency preparedness and risk assessment, (3) birch pollen forecasting, and (4) research and development activities.

The methodology (nesting to country-Denmark and city-scale domains) and several examples of realisation for model-downscaling and integration for urban meteorology and air pollution were recently suggested in the EC FP5 project devoted to Integrated Systems for Forecasting Urban Meteorology, Air Pollution and Population Exposure (EMS-FUMAPEX 2005; see also: http://fumapex.dmi.dk). This way included downscaling from regional (or global) meteorological models to the urban-scale meso-meteorological models with statistically parameterised building effects. Here in this chain further downscaling to the micro-scale with obstacle-resolved Computational Fluid Dynamics (CFD)-type model was suggested. Figure 16.1b demonstrates the above mentioned schemes of the model-downscaling in Urban Air Quality Information and Forecasting Systems (UAQIFS).

16.2 Meteorological Model: HIgh Resolution Limited Area Model (HIRLAM)

The Atmospheric Chemistry–Aerosol–Transport models/modules depend on applicable meteorological driver as well as selected domain of interest; and hence, different meteorological operational or re-analysed archived datasets from NWP (such as DMI-HIRLAM (Unden et al. 2002; Sass et al. 2002) or ECMWF) models (or climate models) can be used. At present, output from several nested versions of DMI-HIRLAM is applied: (Fig. 16.2):

- T15 – 15 × 15 km, 40 vertical layers
- M09 – 9 × 9 km, 40 vertical layers
- S05 – 5 × 5 km, 40 vertical layers

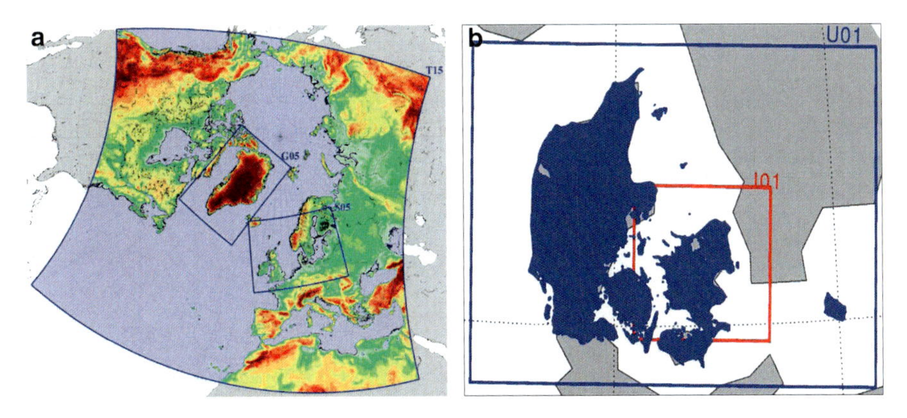

Fig. 16.2 Examples of operational and research NWP DMI-HIRLAM modelling areas

- S03 – 3 × 3 km, 40 vertical layers
- U01/I01 – 1.4 × 1.4 km, 40 vertical layers (experimental urbanised version)

The current operational DMI forecasting modelling system (Yang et al. 2005) includes the pre-processing, climate file generation, data assimilation, initialization, forecasting, post-processing, and verification. It includes also a digital filtering initialization, semi-Lagrangian advection scheme, and a set of physical parameterizations such as Savijaervi radiation, STRACO condensation, CBR turbulence scheme, and ISBA scheme. The lateral boundary conditions are received every 6 h from ECMWF. The system is running on the DMI CRAY-XT5 supercomputer and produced model output files after forecasts is finished, will be archived on a mass storage system.

An interface between the NWP and ACT models was built. Through this interface necessary information is extracted from the HIRLAM output, and then it is used by the ACT model. A comprehensive script system had been built to couple both models together in order to produce air quality forecasts.

16.3 Tropospheric Chemistry–Aerosol–Cloud Modelling System

The Tropospheric Chemistry Aerosol Cloud modelling system is a highly flexible multi-module based system (Fig. 16.3a). There are two versions of the system. The first one (Gross and Baklanov 2004) is a further development of the Lagrangian chemistry model MOON (Gross et al. 2005; Madsen 2006), where the chemistry aerosol cloud module concept makes it easy to perform chemical transformations, and apply with other emission inventories and/or meteorological datasets. Furthermore, the aerosol physics in the new module is more advanced. The second version is Eulerian and moreover, it is based originally on the CAMx model (http://www.camx.com). In the first version the two chemical schemes can be used: the Regional Acid Chemistry Mechanism (RACM) (Stockwell et al. 1997) and an updated version

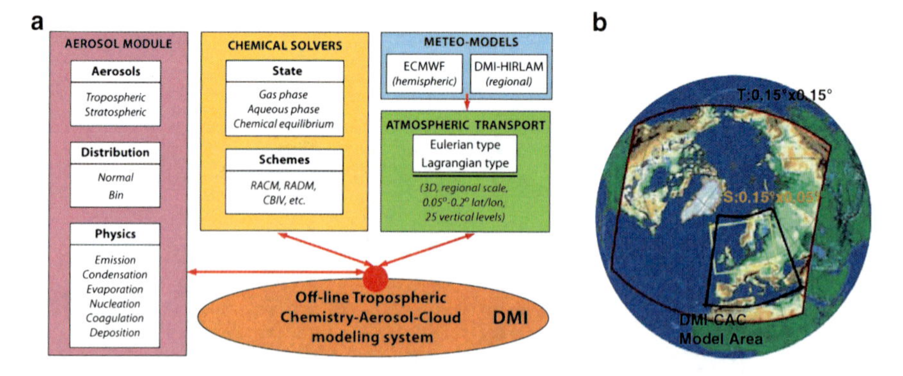

Fig. 16.3 (**a**) Schematic description of the Tropospheric Chemistry Aerosol Cloud modelling systems developed at DMI and (**b**) operational modelling areas

of Carbon Bond IV (CB-IV) Mechanism (Gery et al. 1989) with improved isoprene chemistry. Both mechanisms are used together with the Tropospheric Ultraviolet and Visible radiation model (TUV) (Madronich 2002) to calculate photolysis rate coefficients, and emissions from EMEP (http://www.ceip.at) (Denier 2009).

The aerosol module treats condensation, evaporation, nucleation, deposition and coagulation of aerosols (Baklanov 2002) as shown in Fig. 16.3a. The numerical evolution of aerosols is solved by treating the aerosol size distributions as normal distributions. However, the aerosol module as a part of the modelling system has not been routinely tested yet in the 3D version, i.e. only in 0D (see Gross and Baklanov 2004); and although it was evaluated in the 3D Enviro-HILAM research version of the model (see Korsholm et al. 2008a, b).

The horizontal and vertical resolutions of the model depend on a resolution of the meteorological and emission data. At present the model run over a 0.2° × 0.2° horizontal grid (Fig. 16.3b), and it has a vertical resolution of 25 levels. These vertical levels cover the lowest 3 km of the troposphere. The amount of chemical compounds, which is transported from the free troposphere into the atmospheric boundary layer, is determined by the meteorological information and the concentration of the chemical compounds in the free tropospheric. These concentrations depend on the longitude, latitude, land/sea and month (Gross et al. 2005). The advection is solved using the Bott and PPM schemes (Bott 1989; Colella 1984).

The model system is developed to simulate aerosols and gas-phase compounds from regional to urban scale of ground-level gas-phase air pollutants. It has been used for air quality forecasts of ground-level gas-phase air pollutants and modelling of historical data.

16.4 Enviro-HIRLAM (Environment-HIgh Resolution Limited Area Model)

The Enviro-HIRLAM is an online coupled NWP and ACT model for research and forecasting of both meteorological and chemical weather (Fig. 16.4). The integrated modelling system is developed by DMI and other collaborators[1] (Chenevez et al. 2004; Baklanov et al. 2004, 2008a; Korsholm et al. 2008a, Korsholm 2009) and included by the European HIRLAM consortium as the baseline system in the HIRLAM Chemical Branch (https://hirlam.org/trac/wiki), it is used in several countries.

Enviro-HIRLAM includes two-way feedbacks between air pollutants and meteorological processes. Atmospheric chemical transport equations are implemented inside the meteorological corner on each time step (Chenevez et al. 2004). To make

[1]At the current stage the Enviro-HIRLAM model is used as the baseline system for the HIRLAM chemical branch, and additionally to the HIRLAM community the following groups join the development team: University of Copenhagen, Tartu University (Estonia), Russian State Hydro-Meteorological University and Tomsk State University, Odessa State Environmental University (Ukraine), etc.

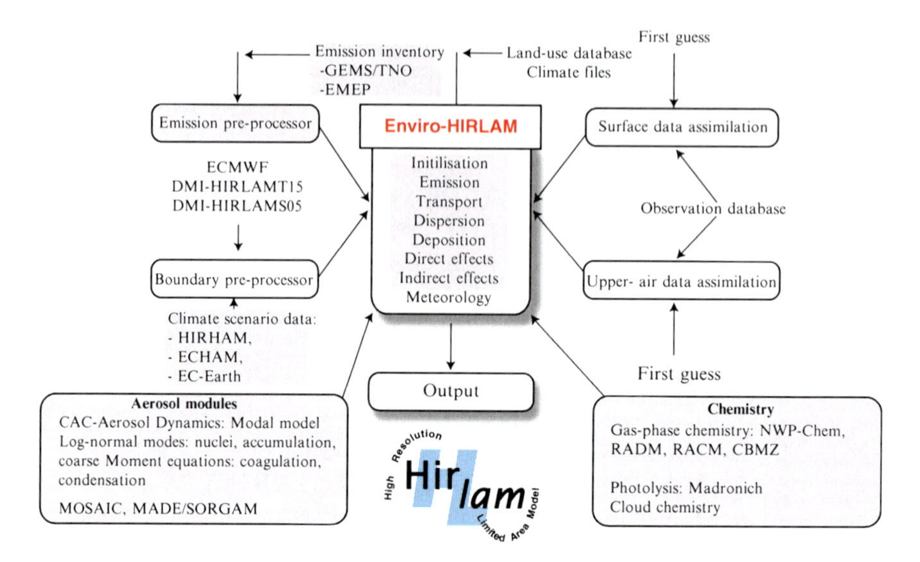

Fig. 16.4 Schematic description of the Integrated (On-line Coupled) Modelling System for Predicting Atmospheric Composition: Enviro-HIRLAM (Environment – HIgh Resolution Limited Area Model)

the model suitable for Chemical Weather Forecast (CWF) in urban areas, where most of population is concentrated, the meteorological part is improved by implementation of urban sublayer modules and parameterisations (Baklanov et al. 2008b). The aerosol module in Enviro-HIRLAM comprises two parts: (1) a thermodynamic equilibrium model (NWP-Chem-Liquid) and (2) the aerosol dynamics model CAC (Gross and Baklanov 2004) based on the modal approach. Parameterisations of the aerosol feedback mechanisms in the Enviro-HIRLAM model are described in Korsholm et al. (2008) and Korsholm (2009). Several chemical mechanisms could be chosen depending on the specific tasks: well-known RADM2 and RACM or new-developed economical NWP-Chem (Korsholm et al. 2008). Validation and sensitivity tests of the on-line versus off-line integrated versions of Enviro-HIRLAM (Korsholm et al. 2009) showed that the online coupling improved the results. Different parts of Enviro-HIRLAM were evaluated versus the ETEX-1 experiment, Chernobyl accident and Paris study datasets and showed that the model performs satisfactorily (Korsholm 2009).

16.5 Danish Emergency Response Model of the Atmosphere (DERMA)

The Danish Emergency Response Model of the Atmosphere (DERMA; Fig. 16.5) is off-line three-dimensional Lagrangian long-range dispersion model using a puff diffusion parameterisation, particle-size dependent deposition parameterisations

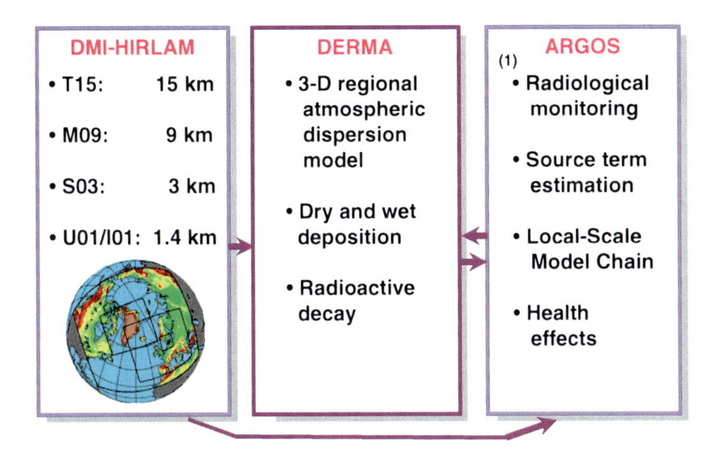

Fig. 16.5 Links between the HIRLAM (NWP models), DERMA, and ARGOS (real-time on-line nuclear decision support system) modules

and radioactive decay (Sørensen 1998; Sørensen et al. 1998, 2007; Baklanov and Sørensen 2001). Earlier comparisons of simulations with the DERMA model versus the ETEX experiment involving passive tracer measurements gave very good results (Graziani et al. 1998). The DERMA model can be used with different sources of NWP data, including the DMI-HIRLAM limited-area and the ECMWF global NWP models with various resolutions. The main objective of DERMA is the prediction of the atmospheric transport, diffusion, deposition and decay of a radioactive plume within a range from about 20 km from the source up to the global scale. DERMA is run on operational computers at DMI. The integration of DERMA in ARGOS is effectuated through automated on-line digital communication and exchange of data. The calculations are carried out in parallel for each NWP model to which DMI has access, thereby providing a mini-ensemble of dispersion forecasts for the emergency management.

16.6 Urban Scale High Resolution Modelling

It is known that the boundary layer in the urban areas has a complex structure due to multiple contributions of different parameters, including variability in roughness and fluxes, etc. All these effects can be included to some extend into models. For research purposes, the DMI–HIRLAM–U01/I01 models (resolution of 1.4 km) with domains shown in Fig. 16.2 are employed for high resolution urbanized modelling (example is shown in Fig. 16.6b). The land-use classification is based on CORINE dataset (http://etc-lusi.eionet.europa.eu/CLC2000) and climate generation files.

The simple urbanization (Fig. 16.6a) of NWP includes modifications (anthropogenic heat flux, roughness, and albedo) the land surface scheme so-called the Interaction Soil Biosphere Atmosphere (ISBA) scheme originally proposed by

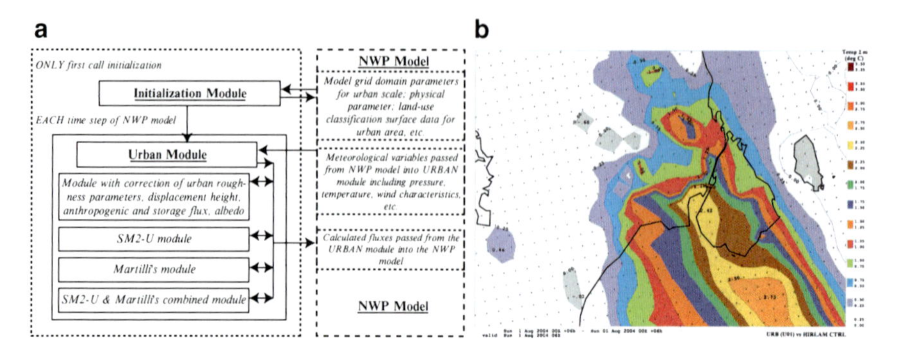

Fig. 16.6 (**a**) General scheme of the NWP model urbanization and (**b**) Difference plots (between outputs of the DMI–HIRLAM control and urbanized runs using AHF+R module – modifications of the ISBA land-surface scheme) for the air temperature at 2 m on 1st August 2004 at 06 UTC

Noilhan and Planton (1989). The changes of the ISBA scheme include modifications of the set of parameters in each grid cells of modelling domain where the urban class is presented. These modifications include the urban roughness, anthropogenic heat flux and albedo. The urban roughness changes up to a maximum of 2 m for grid cells where the urban class will reach up to 100%. The anthropogenic heat flux (from 10 to 200 W/m^2) is modified similarly to the roughness. Albedo is tested for the summer vs. winter cases – from 0.2 to 0.4 by factor of two, i.e. when the snow is covering the surface. Other parameterization (Fig. 16.6a) is the Building Effect Parameterization (BEP) module includes the urban sub-layer parameterization suggested by Martilli et al. (2002) and is used to simulate the effect of buildings on a meso-scale atmospheric flow. It takes into account the main characteristics of the urban environment: (1) the vertical and horizontal surfaces (wall, canyon floor and roofs), (2) the shadowing and radiative trapping effects of the buildings, (3) the anthropogenic heat fluxes through the buildings wall and roof. In this parameterization, the city is represented as a combination of several urban classes. Each class is characterized by an array of buildings of the same width located at the same distance from each other (canyon width), but with different heights (with a certain probability to have a building with specific height). To simplify the formulation the length of the street canyons is assumed equal to the horizontal grid size. The vertical urban structure is defined on a numerical grid.

16.7 Micro-scale Model for Urban Environment (M2UE)

The M2UE (Micro-scale Model for Urban Environment; Nuterman 2008) is Computational Fluid Dynamics (CFD) microscale model for analysis of atmospheric processes and pollution prediction in the urban environment, which takes into account a complex character of aerodynamics in non-uniform urban relief with penetrable (vegetation) and impenetrable (buildings) obstacles and traffic induced

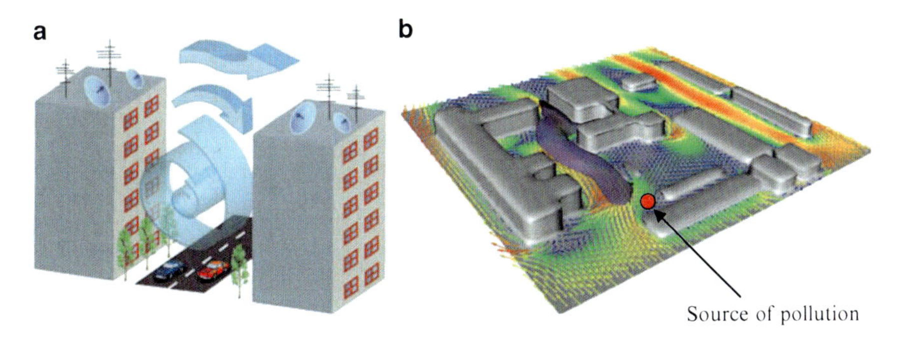

Fig. 16.7 (**a**) Schematic view of typical element of urban canopy – street canyon; (**b**) Wind flow and pollution dispersion for the part of Copenhagen area

turbulence (Fig. 16.7). The model includes steady/unsteady three-dimensional system of Reynolds equations, two-equation k-ε model of turbulence and the 'advection-diffusion' equation to simulate pollution transport. The numerical solution is based on implicit time advancing scheme and finite volume method. M2UE was evaluated by experimental data obtained from the TRAPOS project (Optimization of Modelling Methods for Traffic Pollution in Streets) and COST-732 Action which was devoted to Quality Assurance and Improvement of Microscale Models. In general, the model showed good and realistic results (Nuterman et al. 2008; Baklanov and Nuterman 2009).

16.8 Example of Chemical Weather Forecasting

The Atmospheric Chemistry Transport modelling system used is based on the off-line coupled CAMx and HIRLAM models has been developed to simulate particulate and gas-phase air pollution on different scales. It has been used to simulate short and long-term releases of different chemical species and air pollution episodes. At present it is run in a pre-operational mode 4 times per day based on 3D meteorological fields produced by the HIRLAM NWP model. Currently this modelling system is setup to perform chemical weather forecasts for a series of chemical species (such as O_3, NO, NO_2, CO and SO_2) and forecasted 2D fields at surface are available for each model as well as an ensemble of models (based on 12 European regional air quality models). The simulated output is publicly available and it is placed at the ECMWF website (http://gems.ecmwf.int/d/products/raq/forecasts/) of the EC FP6 GEMS project.

Examples of the surface ozone and nitrogen dioxide forecasts vs. observations are shown in Fig. 16.8. The produced plots are forecasts (top panel) of O_3 and NO_2 at 00 UTC on 19 Mar 2010 together with near-real-time observations (bottom panel) of these species. The simulations resulted in a relatively medium air pollution levels over the Central and Eastern Europe with a few hot-spots. For O_3 it is an episode observed in the east of the Northern Italy, Austria and Slovenia; and

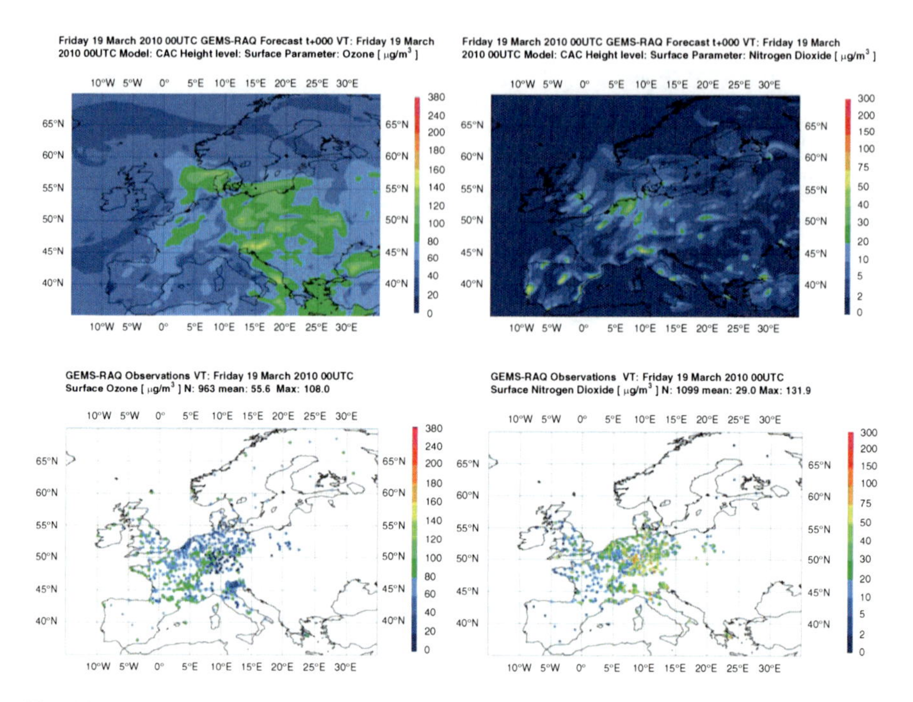

Fig. 16.8 GEMS Regional Air Quality Forecast by the DMI modelling system for ozone and nitrogen dioxide (*top panel*) forecasts vs. (*bottom panel*) observations on 19 Mar 2010, 00 UTC

there are higher concentrations of NO_2 from the metropolitan areas of London, Rein-Ruhr, Paris, Po Valley, Istanbul, and others. In general, there is a good agreement between observations and produced forecasts. Furthermore, a natural correlation and trend are observed between O_3 and NO_2.

Acknowledgement The research leading to these results has received funding from research projects of the EC Programme FP/2007–2011. The authors are thankful to Drs. Bent Sass (from DMI), Allan Gross (formerly at DMI, now at DMU) for discussions and comments. Especial thanks to DMI IT Department for advice and computing support. Authors are thankful to Drs. Vincent-Henri Peuch (Meteo-France) and Miha Razinger (European Centre for Medium-Range Weather Forecasts, ECMWF) for providing access to the GEMS ensemble output dataset. The studies are also a part of the research of the "Center for Energy, Environment and Health", financed by the Danish Strategic Research Program on Sustainable Energy under contract no 2104-06-0027.

References

Baklanov A (2002) Modelling of formation and dynamics of radioactive aerosols in the atmosphere. Res Theory Elem Part Solid State 4:135–148

Baklanov AA, Nuterman RB (2009) Multi-scale atmospheric environment modelling for urban areas. Adv Sci Res 3:53–57

Baklanov A, Sørensen JH (2001) Parameterisation of radionuclide deposition in atmospheric dispersion models. Phys Chem Earth 26:787–799

Baklanov A, Gross A, Sørensen JH (2004) Modelling and forecasting of regional and urban air quality and microclimate. J Comput Technol 9:82–97

Baklanov A, Korsholm U, Mahura A, Petersen C, Gross A (2008a) ENVIRO-HIRLAM: on-line coupled modelling of urban meteorology and air pollution. Adv Sci Res 2:41–46

Baklanov AA, Korsholm U, Nielsen NW, Gross A (2008) On-line coupling of chemistry and aerosols into meteorological models: advantages and prospective, EU FP6 GEMS Integrated Project report, GEMS-VAL deliverable for Task 1.6.3, 25 pp. Available from: http://gems.ecmwf.int/do/get/PublicDocuments/2141

Bott A (1989) A Positive Definite Advection Scheme Obtained by Nonlinear Renormalization of the Advective Fluxes. AMS 117:1006–1015

Chenevez J, Baklanov A, Sorensen JH (2004) Pollutant transport schemes integrated in a numerical weather prediction model: model description and verification results. Meteorol Appl 11:265–275

Colella P, Woodward PR (1984) The Piecewise-Parabolic Method (PPM) for Gas Dynamical Simulations. J. Comput. Phys 54:174–201

Denier van der Gon HAC, Visschedijk AJH, van der Brugh H, Dröge R, and Kuenen J (2009) MEGAPOLI Scientific Report 09–02, A base year (2005) MEGAPOLI European gridded emission inventory (1st version), MEGAPOLI Deliverable 1.2, 17 p, http://megapoli.dmi.dk/publ/MEGAPOLI_sr09-02.pdf

EMS-FUMAPEX (2005) Urban meteorology and atmospheric pollution. Atmos Chem Phys J 24 (Special Issue). Available at http://www.atmos-chem-phys.net/special_issue24.html

Gery MW, Witten GZ, Killus JP, Dodge MC (1989) A photochemical kinetics mechanism for urban and regional scale computer modelling. J Geophys Res 94:925–956

Graziani G, Klug W, Moksa S (1998) Real-time long-range dispersion model evaluation of the ETEX first release. Joint Research Centre, EU, Luxemberg

Gross A, Baklanov A (2004) Modelling the influence of dimethyl sulphide on the aerosol production in the marine boundary layer. Int J Environ Pollut 22(1/2):51–71

Gross A, Sørensen JH, Stockwell WR (2005) A multi-trajectory chemical-transport vectorized gear model: 3-D simulations and model validation. J Atmos Chem 50:211–242

Korsholm U (2009) Integrated modeling of aerosol indirect effects – development and application of a chemical weather model. PhD thesis, University of Copenhagen, Niels Bohr Institute and DMI, Research department. http://www.dmi.dk/dmi/sr09-01.pdf

Korsholm US, Baklanov A, Gross A, Mahura A, Sass BH, Kaas E (2008) Online coupled chemical weather forecasting based on HIRLAM – overview and prospective of Enviro-HIRLAM. HIRLAM Newsl 54:151–168

Korsholm US, Baklanov A, Gross A, Sørensen JH (2009) On the importance of the meteorological coupling interval in dispersion modeling during ETEX-1. Atmos Environ

Madronich S (2002) The prohospheric Visible Ultra-Violet (TUV) model webpage. National Center for Atmospheric Research, Boulder, CO

Madsen MS (2006) Modeling of the Arctic ozone depletion. PhD Thesis, Chemical Institute of University of Copenhagen and Danish Meteorological Institute, Denmark

Mahura A, Baklanov A, Rasmussen A, Korsholm US, Petersen C (2006) Birch pollen forecasting for Denmark. In: 6th Annual meeting of the European Meteorological Society (EMS), Vol 3. Ljubljana, Slovenia, 3–7 Sep 2006, EMS2006-A-00495

Martilli A, Clappier A, Rotach MW (2002) An urban surface exchange parameterisation for mesoscale models. Bound Layer Meteorol 104:261–304

Noilhan J, Planton S (1989) A simple parameterization of land surface processes for meteorological models. Mon Weather Rev 117:536–549

Nuterman R (2008) Modelling of turbulent flow and pollution transport in urban canopy. PhD Thesis, Tomsk State University, 156 p

Nuterman RB, Starchenko AV, Baklanov AA (2008) Development and evaluation of a microscale meteorological model for investigation of airflows in urban terrain. J Comput Technol 13(3):37–43

Rasmussen A, Mahura A, Baklanov A, Sommer J (2006) The Danish operational pollen forecasting system. 8th International congress on aerobiology, towards a comprehensive vision, Neuchâtel, Switzerland, 21–25 Aug 2006

Sass BH, Nielsen NW, Jørgensen JU, Amstrup B, Kmit M, Mogensen KS (2002) The Operational DMI-HIRLAM System-2002-version, DMI Tech. Rep., 99-21

Sørensen JH (1998) Sensitivity of the DERMA long-range dispersion model to meteorological input and diffusion parameters. Atmos Environ 32:4195–4206

Sørensen JH, Rasmussen A, Ellermann T, Lyck E (1998) Mesoscale influence on long-range transport; evidence from ETEX modelling and observations. Atmos Environ 32:4207–4217

Sørensen JH, Baklanov A, Hoe S (2007) The Danish emergency response model of the atmosphere (DERMA). J Environ Radioact 96:122–129

Stockwell WR, Kirchner F, Kuhn M (1997) A new mechanism for regional atmospheric chemistry modeling. Journal of Geophysical Research 102(D22): 25847–25879

Unden P, Rontu L, Järvinen H, Lynch P, Calvo J, Cats G, Cuhart J, Eerola K, etc. (2002) HIRLAM-5 Scientific Documentation. December 2002, HIRLAM-5 Project Report, SMHI

Yang X, Petersen C, Amstrup B, Andersen B, Feddersen H, Kmit M, Korsholm U, Lindberg K, Mogensen K, Sass B, Sattler K, Nielsen W (2005) The MDI-HIRLAM upgrade in June 2004. DMI Technical Report, 05–09, 35 p

Part III
Validation and Case Studies

Chapter 17
Chemical Modelling with CHASER and WRF/Chem in Japan

Masayuki Takigawa, M. Niwano, H. Akimoto, and M. Takahashi

17.1 Introduction

Recently, chemical transport models (CTMs) have enhanced flight planning by providing direct information on the expected state of important 3D atmospheric chemical structures on timescales from hours to days, i.e., the "chemical weather". The first chemical weather forecasts (CWFs) was performed during the Airborne Southern Hemisphere Ozone Experiment (ASHORE) and the Second European Stratospheric Arctic and Middle-Latitude Experiment (SESAME) in 1994 and 1995, respectively (Lee et al. 1997). The use of CWFs for field campaigns is expanding rapidly. Global CWFs from the Model of Atmospheric Transport and Chemistry (MATCH; of the Max-Planck-Institute for Chemistry, MPIC) were used during the Indian Ocean Experiment (INDOEX) in 1999, and the Mediterranean Intensive Oxidants Study (MINOS) and the Convective Transport of Trace Gases into the Upper Troposphere over Europe (CONTRACE) in 2001 (Lawrence et al. 2003). In contrast to regional CWFs, global CWFs can predict intercontinental transport. For example, Lawrence et al. (2003) estimated the frequency of intercontinental pollution plumes from North America and Asia to Europe. Chemical weather forecast systems using regional scale models have an advantage – a higher horizontal resolution – compared with global models. The Chemical Weather Forecasting System (CFORS) was used during the Transport and Chemical Evolution over the Pacific (TRACE-P) and the Asian Pacific Regional Aerosol Characterization Experiment (ACE-Asia) campaign in 2001 (Uno et al. 2003). Uno et al. (2003) showed that changes in synoptic-scale weather patterns greatly influence continental-scale pollution transport in spring over East Asia.

This paper describes a newly developed global CWF system based on CHASER model that can be used to support atmospheric chemistry field campaigns. The model includes radiative and dynamical processes, and comprehensive chemical

M. Takigawa (✉)
Japan Agency for Marine-Earth Science and Technology, 3173-25 Showa-machi, Kanazawa-ku, Yokohama, Kanagawa 236-0001, Japan
e-mail: takigawa@jamstec.go.jp

A. Baklanov et al. (eds.), *Integrated Systems of Meso-Meteorological and Chemical Transport Models*, DOI 10.1007/978-3-642-13980-2_17,
© Springer-Verlag Berlin Heidelberg 2011

schemes for the troposphere and lower stratosphere. This paper assesses the quality and estimates the value of CWFs from the model that was used in flight campaigns.

17.2 Global Chemical Weather Forecasting System

The chemical weather forecast (CWF) system includes the coupled tropospheric chemistry climate model CHASER, which is described and evaluated in Sudo et al. (2002a, b). Physical and dynamical processes are simulated following the Center for Climate System Research/National Institute for Environmental Study/ Frontier Research Center for Global Change (CCSR/NIES/FRCGC) atmospheric GCM (Nakajima et al. 1995; Numaguti 1993; Numaguti et al. 1995). In this study the CHASER model is based on CCSR/NIES/FRCGC AGCM version 5.7b. Advective transport is simulated with a 4-th order flux-form advection scheme using a monotonic Piecewise Parabolic Method (PPM) (Colella and Woodward 1984) and a flux-form semi-Lagrangian scheme (Lin and Rood 1996). Subgrid-scale vertical fluxes of heat, moisture, and tracers are approximated using a non-local turbulence closure scheme based on Holslag and Boville (1993) used in conjunction with the level 2 scheme of Mellor and Yamada (1974). The cumulus parameterization scheme is based on Arakawa and Schubert (1974) with several simplifications described in Numaguti et al. (1997). The closure assumption is changed from the diagnostic closure used in Numaguti et al. (1997) to a prognostic closure based on Pan and Randall (1998), in which cloud base mass flux is treated as a prognostic variable. An empirical cumulus suppression condition introduced in Emori et al. (2001) is adopted. Note that both the updraft and downdraft of chemical species by cumulus convection are included in the model. The large-scale condensation scheme is based on Treut and Li (1991), in which subgrid probability distribution of total water mixing ratio in each grid box is assumed to have a uniform distribution. Spectral coefficients are triangularly truncated at wavenumber 42 (T42), equivalent to a horizontal grid spacing of about 2.8°. The model has 32 vertical layers that are spaced at about 1 km intervals in the free troposphere and lower stratosphere. The chemical side of the model is based on Sudo et al. (2002a, 2003), and includes a detailed on-line simulation of tropospheric chemistry involving the O_3–$HO\mathit{x}$–$NO\mathit{x}$–CH_4–CO system and oxidation of NMHCs. The chemical model time step is 10 min. The model includes detailed dry and wet deposition schemes and heterogeneous reactions on the surface of sulfate and nitrate aerosols.

In addition to the extensive chemical reactions, forecast runs include CO tracers which are "tagged" with their origin. Such tracers are emitted normally over selected regions (north and south China, Japan, south Asia, northern America, Europe, and Siberia) and evolved subject to model transport schemes and normal chemical loss processes for CO. Anthropogenic surface emissions of CO are taken from the Streets et al. (2003) inventory over Asia (except China), and from the Emission Database for Global Atmospheric Research (EDGAR) (Olivier et al. 1996) over other regions. Surface CO emissions over China are taken from D. Streets, personal communication. The estimated annual amount of emissions over China is 146 TgCO, a figure about 40 TgCO larger than that in Streets et al. (2003). In this study, the timing of CO emission from biomass burning was estimated by using the average of the hot spot data from 1995 to 2001 from Along Track Scanning Radiometer (ATSR) (Arino et al. 1999) for the daily forecasts. ATSR hot spot data for 2002 were used in the post-analysis study. Tagged CO tracer is also considered for CO which is chemically produced from the oxidation of CH_4, isoprenes, and other NMHCs.

Each daily run is fully automated and consists of two parts: a "quasi-real-time" run and forecast run. The quasi-real-time run is derived from the NCEP final analysis (FNL) data, and steps forward 1 day at a time as soon as the previous day's data are available. Forecast runs use the NCEP Global Forecast System (GFS) data instead of the NCEP FNL data. The NCEP data are re-gridded from $1^\circ \times 1^\circ$ to $2.8^\circ \times 2.8^\circ$ in horizontal, and from 24 to 32 layers in vertical. The relaxation time for nudging the CHASER model meteorological field to the NCEP meteorological data is 1 day in the free troposphere and lower stratosphere. The relaxation time approaches 0 at the surface. Sea surface temperature (SST) is based on the WMO Distributed Data Bases managed by the Japan Meteorological Agency. Winds from 10 to 3 hPa are calculated by using dynamical and physical procedures in the CHASER model. They are not nudged by NCEP data because the maximum height of NCEP data is lower than the top of the CHASER model. Humidity is calculated using the hydrological cycle (surface sources and sinks, transport, convection, diffusion, condensation, and precipitation). Consequently, the temperature and humidity fields do not produce destabilization or discontinuity, still there might be inconsistencies between the CHASER temperature and NCEP humidity. The forecast run is initialized from a restart file written at the end of the previous day's quasi-real-time run. Automated runs normally start at 05:00 of the local (Japanese) standard time, JST (i.e. 20:00 UTC of the previous day). Therefore, the 1 day forecast was available for pre-flight briefing during the PEACE campaign. Pre-formatted figures are automatically made from output from both the quasi-real-time and forecast runs. Archived output data can be used to make custom figures via the web interface (http://www.jamstec.go.jp/frcgc/gcwm).

This forecast system evolved from a prototype run with lower resolution (T21) that started in November 2000. Then it switched to higher resolution (T42) with a spin-up time of 3 months for the global distribution of chemical species. Daily forecasts and archived output data have been available on a web page at the Japan Agency for Marine-Earth Science and Technology (JAMSTEC) since 1 January 2002.

17.3 Results

The climatological and seasonal distributions of chemical species (i.e., the chemical climate) calculated by CHASER have already been evaluated in Sudo et al. (2002b, 2003). The focus here is on the validation and interpretation of the chemical structures on timescales from hours to days. The targeted features of the flights based on CWFs can be characterized by: a) pollution plumes that were affected by intercontinental transport, and b) the outflow of a polluted air mass from nearby populated/industrial regions. The global CWF can globally predict to where the polluted air will be transported on a day-to-day bias. However, such predictions would benefit from global or regional model forecasts with higher resolution, especially, if they are used to plan flight paths.

17.3.1 Meteorological Fields

An accurate characterization of transport processes is of critical importance to flight planning and analysis of observations. Figure 17.1 compares PEACE G-II observations of meteorological parameters (wind direction and speed, air temperature, and

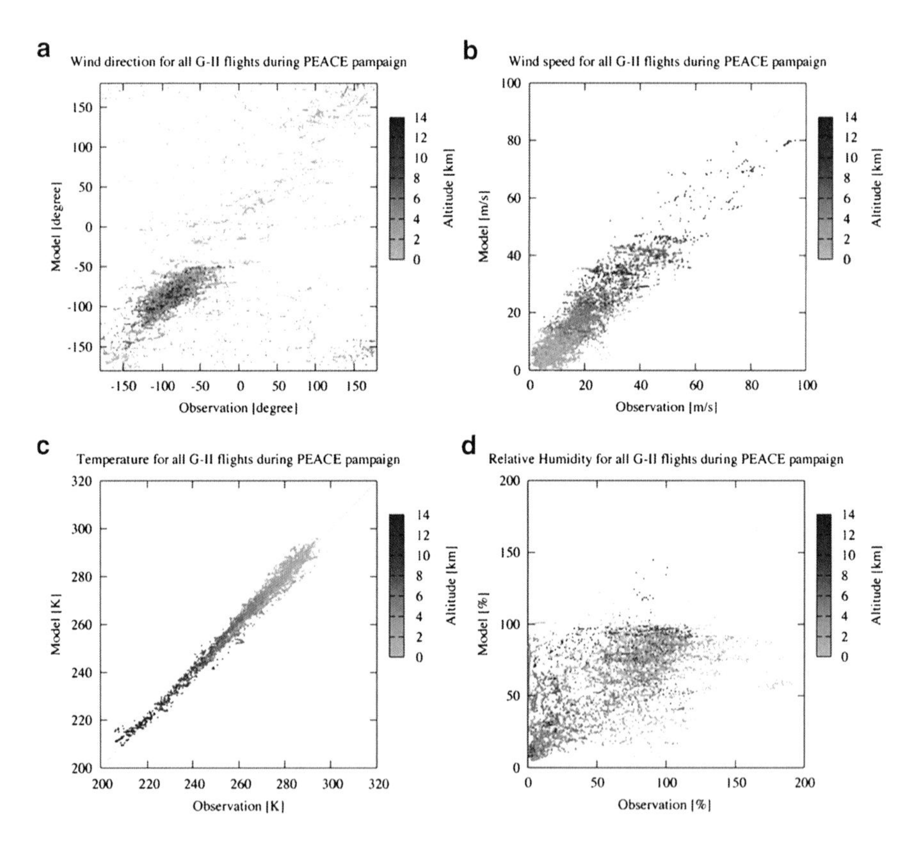

Fig. 17.1 Comparison between PEACE G-II airborne meteorological parameters and global chemical weather forecasting system output fields along the flight paths. Colors denote altitudes

relative humidity) with CHASER simulated meteorological output. Model output was extracted along the G-II aircraft flight path and compared to aircraft observations averaged over 60 s. The color of each dot denotes the height. Wind speed, and air temperature well agreed with the G-II measurements. The model tends to overestimate relative humidity in the low-humidity regions. In addition, the model did not reproduce the supersaturated air seen in flight 12 of PEACE-A on 21 January and flight 11 of PEACE-B on 15 May. The flight paths on these dates passed through a low-pressure system, and because of a coarse resolution the model failed to resolve fine structures. By the same reason, the model could not reproduce the wind direction in the lower troposphere (red dots in Fig. 17.1), especially in PEACE-B. The correlation coefficients between modelled and observational data for all G-II flights were 0.81, 0.91, 0.99, and 0.66 for the wind direction, wind speed, temperature, and relative humidity, respectively. The CHASER global CWF accurately captured many of important observed meteorological features during PEACE. The results presented are driven by the NCEP FNL data. The forecast and hindcast meteorological fields were similar during both the PEACE-A and PEACE-B campaigns.

17.3.2 Comparison with Ground-Based Observations

The mixing ratios of chemical species calculated by the CWF system during PEACE were compared with values observed at three ground-based observational sites in the major sampling region. Figure 17.2 shows CO at Minamitorishima (24.2°N, 153.6°E), Yonagunijima (24.3°N, 123.1°E), and Ryori (39.2°N, 141.5°E). The model reproduced the observed temporal variations of each site except small peaks at Yonagunijima, especially in winter. Although biomass-burning emissions based on the 2001–2002 ATSR hot spot data were incorporated into the model, the model still underestimated CO mixing ratio by about 20 ppbv at these sites in spring. The hot spot data are well correlated with emission anomalies, especially for Siberia (Yurganov et al. 2005).

Figure 17.2 shows the regional CO tracers calculated by CHASER. Siberian CO made small contribution to the CO concentrations in late spring at these sites. The concentration of Siberian CO does not exceed 5 ppbv in late spring. Minamitorishima is in the southeast of Japan and is affected by maritime air and outflow from Asia. Tagged CO tracers suggest that the enhanced CO level observed at Minamitorishima on 18 January is linked to emissions from northern and southern China. The CO level at Yonagunijima is strongly related to the Asian CO tracers. Some events with increased CO levels (greater than 300 ppbv) observed at Yonagunijima. Corresponding increases occurred in the China CO tracers in such events. Tracers of CO that is chemically produced from the oxidation of CH_4, isoprenes, and other NMHCs, increased by about 5–10 ppbv in the CO elevated event in the spring. Increases in CO tracers, linked to chemical production from hydrocarbons in polluted air masses, reflect the enhanced chemical activity in the spring.

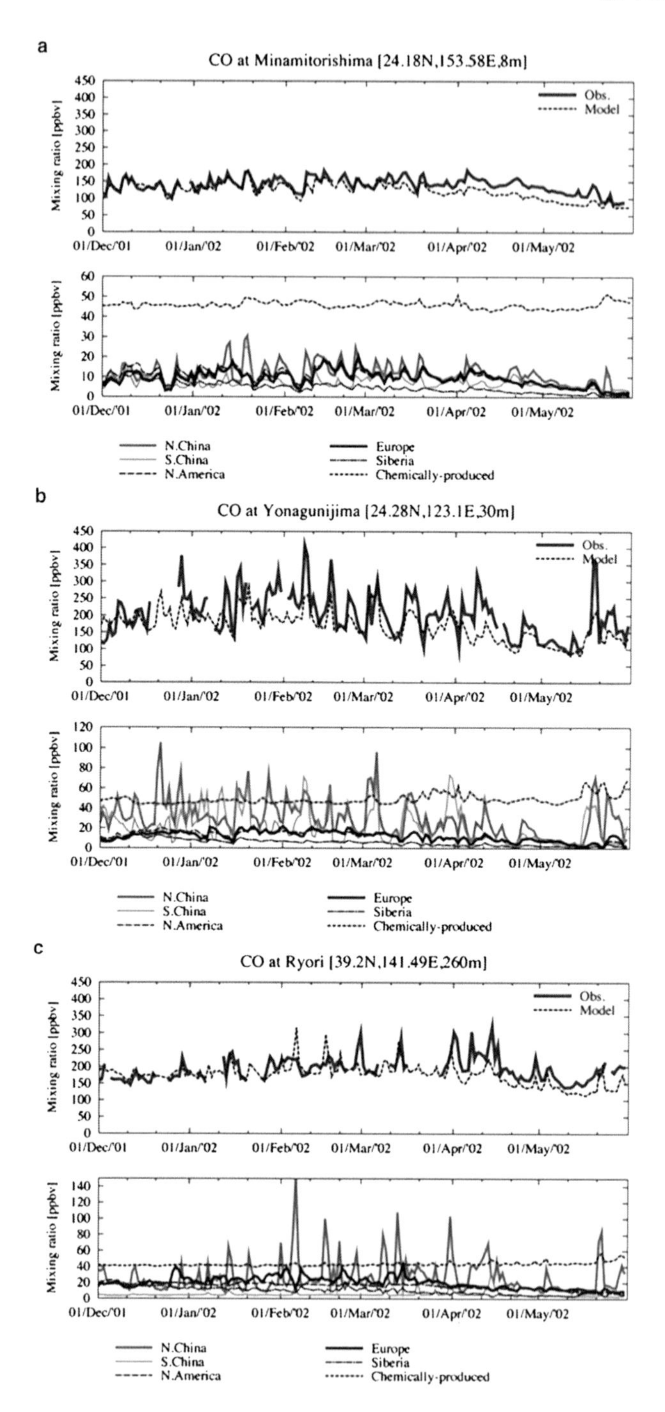

Fig. 17.2 Observed (*bold solid lines*) and modeled (*dotted lines*) surface CO mixing ratio at measurement stations: (**a**) Minamitorishima (24.2°N, 153.6°E), (**b**) Yonagunijima (24.3°N, 123.1°E), and (**c**) Ryori (39.2°N, 141.5°E). Regional tagged CO tracers for northern China and Korea, southern China, south Asia, Europe, North America, and chemically production (*dotted lines*) are also shown

17.3.3 Convective Outflow During PEACE-B Campaign

Convective activity can play an important role in late spring, when PEACE-B was conducted. Oshima et al. (2004) evaluated the origin of air parcels sampled by the aircraft during PEACE-B from altitudes between 4 and 13 km using backward trajectories and estimated that 69% of those that originated at/or below 800 hPa experienced convective uplifting. CCSR/NIES/FRCGC AGCM and CHASER consider tracer updraft and downdraft corresponding to deep cumulus convections. Figure 17.3 shows the modeled mass change of CO tracers in the free troposphere (above 2 km) caused by convective transport over northern China (poleward of 30°N) and Korea, southern China (equatorward of 30°N), and Japan. The model results show greater vertical transport resulting from deep convection over China in late spring. A clear increase also occurred in January over Japan. It is related to passages of mid-latitude cyclones over Japan. The monthly budget of convective transport for May is estimated to be about 3 and 3.3 TgCO over northern and southern China, respectively. These values are about half of all the surface emissions over China in this month. The monthly budgets of surface emissions in the model over these regions are about 6.5 and 6.4 TgCO, respectively.

Several PEACE-B flights were carried out in May 2002 when convective CO transport was active. Figure 17.4 shows the vertical profiles of the observed and modeled CO mixing ratios on 14 May 2002 (flight 10, PEACE-B). CO increased at levels between 300–400 hPa. The model typically underestimated background CO

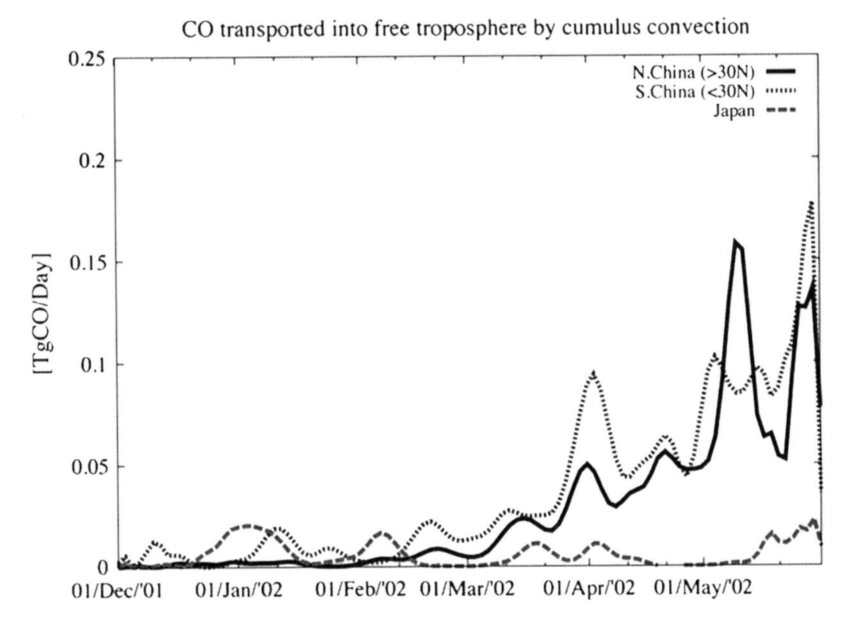

Fig. 17.3 Ten-day running mean of CO transport by convection over northern China (*solid line*), southern China (*dotted line*), and Japan (*dashed line*) during PEACE as calculated by CHASER

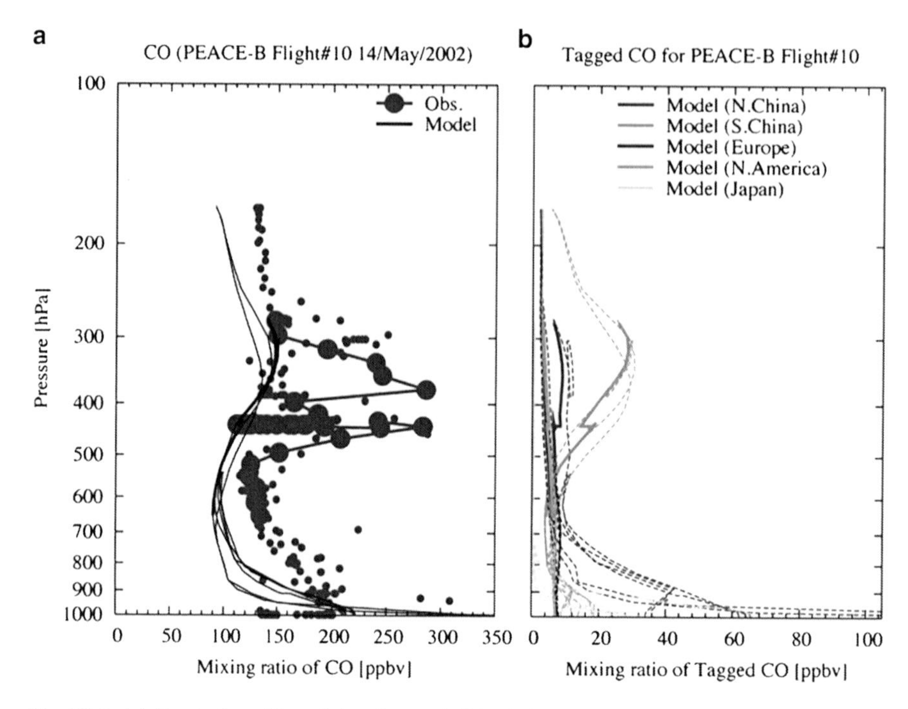

Fig. 17.4 (**a**) Vertical profiles of the observed CO mixing ratios (*gray circles*), along with the model output corresponding to the flight tracks for PEACE-B flight 10 on 14 May 2002. *Black line* denotes the total CO calculated by CHASER. (**b**) Regional tagged CO tracers for northern China and Korea, southern China, Europe, North America, Siberia, Japan, and biomass-burning outside from Asia. The *bold lines* indicate CO values from 06:48 UTC to 07:22 UTC, respectively

level by about 20–40 ppbv compared to the PEACE-B observations, but for this flight, the model showed similar UT CO enhancement. The correlation coefficient between the observed and calculated ΔCO for this flight is 0.56 in the FT. The observed CO concentration has increased from about 120 ppbv (outside the plume) to around 300 ppbv (inside the plume). There is a similar increase in the modeled CO profile, although the modeled CO increase is smaller (i.e., from 80 to 145 ppbv), and it is restricted to 300–350 hPa. Tagged CO tracers suggest that emissions from southern China are responsible for this enhanced UT plume. Emissions from southern and northern China contribute 18–21 and 8–10%, respectively, of the total CO mixing ratio in the plume. Figure 17.5 shows the sea level pressure, and the CO mixing ratio tendency forced by convective transport, CO fluxes, and modeled CO mixing ratio (all given at 300 hPa) calculated by CHASER.

Figure 17.5 also shows the track of flight 10 during PEACE-B on 14 May 2002. A mid-latitude cyclone was located over central China on 13 May. Warm and moist southerly winds in the lower troposphere intensified ahead of this surface cyclone over southern China, converging into a front extending between 20°N, 100°E and 30°N, 115°E. Deep convective clouds, with tops above 10 km around 30°N, 105–120°E, are found in an infrared (IR) image obtained by the Geostationary

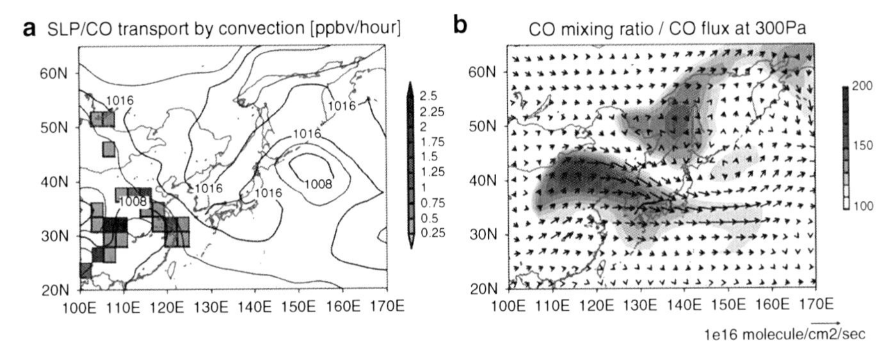

Fig. 17.5 (**a**) Sea level pressure (contours, units of hPa), and CO tendency by convective transport at 300 hPa at 18:00 UTC on 13 May 2002 (*gray tones*, units of ppbv/hour), (**b**) CO mixing ratio and horizontal CO flux at 300 hPa at 18:00 UTC on 13 May 2002. The contour interval is 50 ppbv. The unit length is shown below the figure. *Black lines* denote the track of flight 10 of PEACE-B on 14 May 2002, and the leg from 06:48 UTC to 07:22 UTC is shown with the *bold line*

Meteorological Satellite (GMS)-5 (see Fig. 6d of Oshima et al. (2004)). CO over southern Asia in the UT was transported by southerly LT winds over southern China, and trapped by the cumulus convection in that area. The model reproduces similar convective activity around the cyclone and models the CO changes influenced by convective transport, as shown in Fig. 17.5a. Convective transport changes the CO concentration at an estimated rate of 0.5–3 ppbv/h at 300 hPa over central China. The mass change of CO in the free troposphere caused by the uplift of cumulus convection was calculated by using cloud base mass flux and detraining mass flux in the model. The mass change of CO by the downdraft is also considered, but the effect is small compared to the uplift over central China. The convective processes associated with this cyclone began on 12 May 2002. Consequently, convection transported about 0.3 TgCO on 12–13 May 2002, and the maximum height of the CO concentration increase caused by convective transport is between 300–400 hPa. A high CO region with the value of 150 ppbv or higher can be seen over northern China and Korea in Fig. 17.5b. The modeled CO increase by convective transport has not been seen at 450 hPa over central China. The model tends to underestimate the detrainment of middle convection, and it is the probable reason why the model was not able to reproduce the CO increase at 450 hPa during the flight. Five–day backward trajectories for air parcels in which the highest CO mixing ratios were observed during this flight indicate that these high-CO air parcels were located over central China (around 30°N, 115°E) 24 h prior to the measurement (see Fig. 5a, b of Oshima et al. (2004)). The CO over central China between levels of 300–400 hPa increased from 0.98 Tg on 11 May to 1.28 Tg on 13 May. The plume over central China was transported from the surface by the cumulus convection during 12–13 May. It was subsequently advected by westerly winds and moved over central Japan on 13–14 May.

17.4 Summary and Conclusions

Global chemical weather forecasts made with CHASER can be useful in planning observational flights targeting different types of synoptic scale phenomena, such as near-surface outflow from nearby polluted regions, or intercontinental plumes of pollution in the middle and upper troposphere. The chemical forecast modelled was able to reproduce the CO values observed during the PEACE-A and PEACE-B campaigns. The values were within 10–20% of the observed mixing ratios at three ground-based observational sites in the major sampling region of PEACE, and within 20–30% of the airborne observed mean mixing ratios. The ability to reproduce spatial and temporal variability is critical for planning of measurement flights. Although the model underestimates the background CO level by 20–40 ppbv compared with observations in late spring, it is still capable to reproduce the transport of polluted air masses in the free troposphere. The model estimated the convective transport of CO. The results suggest that about a half of the emissions over China are affected by the cumulus convection in late spring.

Acknowledgements We thank G. Grell, K. Sudo, and all others responsible for the development of the WRF/Chem and CHASER models. We also thank those responsible for observations at air quality monitoring stations. This study was supported by an internal special project fund of the Japan Agency for Marine-Earth Science and Technology (JAMSTEC).

Appendix A: One-Way Nested Global Regional Model Based on CHASER and WRF/Chem

We are now also developing a one-way nested global-regional air quality forecasting (AQF) model system with full chemistry based on the CHASER (Sudo et al. 2002a) and WRF/Chem (Grell et al. 2005). Here, description and evaluation of model system are briefly given.

The global chemistry transport model (CTM) part is based on the CHASER model, which is based on CCSR/NIES/FRCGC atmospheric general circulation model (AGCM) version 5.7b. The basic features of the model have been already described in Sect. 2. The regional CTM part is based on WRF/Chem (Grell et al. 2005). The databases used are the following:

- Anthropogenic emission data over Japan, except those from automobiles, are from the JCAP (Japan Clean Air Program) with 1×1 km resolution (Kannari et al. 2007)
- Anthropogenic emissions from automobiles over Japan are from EAgrid2000 (East Asian Air Pollutant Emissions Grid Inventory) with 1×1 km resolution (K. Murano, personal communication)

- Surface emissions over China and North and South Korea are from REAS (Regional Emission Inventory in Asia) with $0.5 \times 0.5°$ resolution (Ohara et al. 2007)
- Surface emissions over Russia are from EDGAR (Emission Database for Global Atmospheric Research) with $1 \times 1°$ resolution (Olivier et al. 1996)

Diurnal and seasonal variations in surface emissions are taken into account in the JCAP and EAgrid2000 data, and diurnal variations are also parameterized in emissions from REAS and EDGAR following averaged variations of JCAP. Weekly variation between workdays and holidays is also taken into account in the EAgrid2000 automobile emission data. Note here that emissions based on the statistics in 2000 are applied in the present study. Biogenic emissions are based on Guenther et al. (1993). The outer domain covers Japan with 15 km horizontal resolution (152×52 grids for chemical species), and the inner domain covers the Kanto region with 5 km resolution (111×111 grids for chemical species). The inner and outer domains in the regional CTM have 31 vertical layers (up to 100 hPa). The two-way nesting calculation is applied in the regional CTM part.

The lateral boundary of chemical species in the regional CTM is taken from the global CTM. The output of the global CTM is linearly interpolated from the Gaussian latitude and longitude grid to a Lambert conformal conic projection for use in the regional CTM. The lateral boundary is updated every 3 h and linearly interpolated for each time step. We did not include feedback from the regional CTM to the global CTM; that is, the one-way nesting calculation was done between the global and regional CTMs. The system is driven by meteorological data from NCEP for the global CTM part and from the mesoscale model (MSM) of the Japan Meteorological Agency (JMA) for the regional CTM part. A 15-h forecast has been produced four times daily at 00, 06, 12, and 18 UTC with a lead time of 8–10 h since July 2006, following a spin-up of 1 month for the global distribution of chemical species. The initial condition of the meteorological field for the regional CTM was taken from the MSM for each forecast, and the initial condition of chemical species was taken from the model output driven by the analysis meteorology.

To evaluate the model-calculated ozone, the surface ozone mixing ratio was compared to that observed at air quality monitoring stations. There are 251 stations observing surface ozone within the inner domain of the regional CTM as of August 2006. For the comparison of temporal variation, hourly averaged values of observed and modeled surface O_3 mixing ratios in August 2006 are shown in Fig. 17.6. Observed ozone exceeded 100 ppbv from 3 to 6 August at Hanyuu in Saitama Prefecture (36°10′28″N, 139°33′21″E, Fig. 17.6a), which is downwind of the Tokyo metropolitan area. The maximum value in the observation was 162 ppbv at 16 UTC on 3 August. The model successfully reproduced the ozone maximum on 3 August. The maximum simulated value was 137 ppbv in the model. The model also successfully captured the decrease from 3 to 7 August, but failed to show the rapid decrease on 8 August. Three typhoons (Maria, Somai, and Bopha) occurred during this period, and the difficulty of predicting the meteorological field may have led to the overestimation of ozone on 8 August. Both the model and observations

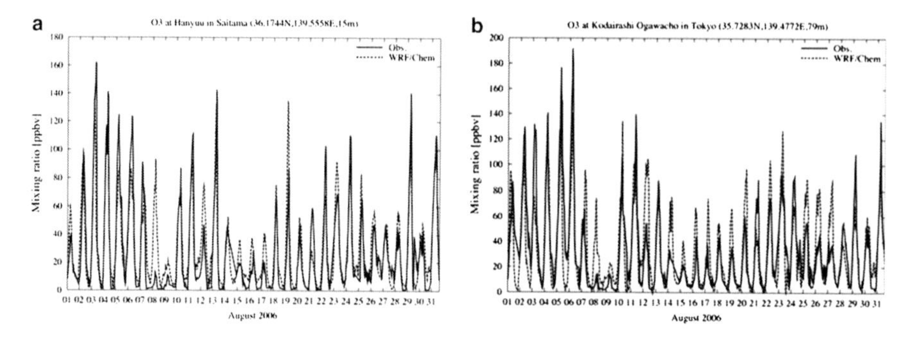

Fig. 17.6 Hourly observed (*solid*) and modelled (*dashed*) surface ozone mixing ratio in August 2006 at Hanyuu in Saitama prefecture (**a**) and Kodaira in Tokyo (**b**)

indicate low levels of ozone from 14 to 17 August as typhoon 200610 (Wukong) approached Japan. The observed and modeled ozone exceeded 100 ppbv on 11 and 13 August, and the model overestimated the ozone mixing ratio on 19 August. The modeled ozone mixing ratio was 135 ppbv, whereas the observed ozone mixing ratio was 86 ppbv. Daily variation in the ozone mixing ratio at nighttime was well reproduced by the model. The daily minimum of observed and modeled ozone exceeded 10 ppbv on 12, 15, 27, and 28 August; except for these days, the ozone level was almost zero during nighttime. The comparison between the modeled and observed daily variation in surface ozone at Kodaira in Tokyo (35°43′42″N, 139°28′38″) is shown in Fig. 17.6b.

Maxima of observed and modeled surface ozone at Hanyuu appeared on 3 August, and the observed and modeled ozone mixing ratios at Kodaira were 140 ppbv or higher on 5 and 6 August. The model tended to overestimate the daytime ozone maximum especially for cloudy days, and the discrepancy of daily maximum is larger in urban area compared to that in rural area. To evaluate the model performance, a set of statistical measures provided by the U.S. Environmental Protection Agency (US EPA 1991) was evaluated for stations in the inner domain of the model. The mean normalized bias error (MNBE), the mean normalized gross error (MNGE), and the unpaired peak prediction accuracy (UPA) were 7.1, 9.5, and 9.4%, respectively. These values are within the criteria range suggested by the U.S. EPA (MNBE$< \pm 10$–15%, MNGE$< \pm 30$–35%, and UPA$< \pm 15$–20%).

References

Arakawa A, Schubert W (1974) Interactions of cumulus cloud ensemble with the large-scale environment. Part I. J Atmos Sci 31:671–701

Arino O, Rosaz J-M, Melinotte J-M (1999) World Fire Atlas with AVHRR and ATSR. Paper presented at IUFRO Conference on Remote Sensing and Forest Monitoring

Colella P, Woodward P (1984) The Piecewise Parabolic Method PPM for gas-dynamic simulations. J Comput Phys 54:174–201

Emori S, Nozawa T, Numaguti A, Uno I (2001) Importance of cumulus parametrization for precipitation simulation over East Asia in June. J Meteorol Soc Japan 79:939–947
Grell GA, Peckham SE, Schmitz R et al (2005) Fully coupled "online" chemistry within the WRF model. Atmos Environ 39:6957–6975
Guenther A, Zimmerman PR, Harley P et al (1993) Isoprene and monoterepene emission rate variability: Model evaluations and sensitivity analyses. J Geophys Res 98D:12609–12617
Holslag A, Boville B (1993) Local versus nonlocal boundary-layer diffusion in a global climate model. J. Climate 6:1825–1842
Kannari A, Tonooka Y, Bada T et al (2007) Development of multiple-species 1 km × 1 km resolution hourly basis emissions inventory for Japan. Atmos Environ 41:3428–3439. doi:10.1016/j.atmosenv.2006.12.015
Lawrence M, Rasch P, Kuhlmann R et al (2003) Global chemical weather forecasts for field campaign planning: predictions and observations of large-scale features during MINOS, CONTRACE, and INDOEX. Atmos Chem Phys 3:267–289
Lee A, Carver G, Chipperfield M, Pyle J (1997) Three-dimensional chemical forecasting: a methodology. J Geophys Res 102:3905–3919
Lin S-J, Rood R (1996) Multidimensional flux-form semi-Lagrangian transport schemes. Mon Weather Rev 124:2046–2070
Mellor GL, Yamada T (1974) A hierarchy of turbulence closure models for planetary boundary layers. J Atmos Sci 31:1791–1806
Nakajima T, Tsukamoto M, Tsusima Y, Numaguti A (1995) Modelling of the radiative process in a AGCM. In: Reports of a new program for creative basic research studies, studies of global environment change to Asia and Pacific regions. Rep. I-3, CCSR, Tokyo, pp 104–123
Numaguti A (1993) Dynamics and energy balance of the Hadley circulation and the tropical precipitation zones: significance of the distribution of evaporation. J Atmos Sci 50:1874–1887
Numaguti A, Takahashi M, Nakajima T, Sumi A (1995) Development of an atmospheric general circulation model. In: Reports of a new program for creative basic research studies, studies of global environment change to Asia and Pacific regions. Rep. I-3, CCSR, Tokyo, pp 1–27
Numaguti A, Takahashi M, Nakajima T, Sumi A (1997) Development of CCSR/NIES atmospheric general circulation model, no. 3 in CGER's Supercomput. Monogr. Rep., CGER, Tsukuba, Ibaraki, pp 1–48
Ohara T, Akimoto H, Kurokawa J et al (2007) Asian emission inventory for anthropogenic emission sources during the period 1980–2020. Atmos Chem Phys 7:4194–4444. doi:www.atmos-chem-phys.net/7/4419/2007
Olivier JGJ, Bouwman AF, Van der Maas CWM et al. (1996) Description of EDGAR Version 2.0. A set of global emission inventories of greenhouse gases and ozone-depleting substances for all anthropogenic and most natural sources on a per country basis and on 1° × 1° grid. RIVM/TNO rep., RIVM, Bilthoven, number nr. 711060 002, 1006
Oshima N et al (2004) Asian chemical outflow to the Pacific in late spring observed during the PEACE-B aircraft mission. J Geophys Res 109, D23S05, doi:10.1029/2004JD004976.
Pan D-M, Randall D (1998) A cumulus parametrization with a prognostic closure. Q J R Meteorol Soc 124:949–981
Streets D et al (2003) An inventory of gaseous and primary aerosol in Asia in the year 2000. J Geophys Res 108(D21):8809. doi:10.1029/2002JD003093
Sudo K, Takahashi M, Kurokawa J, Akimoto H (2002a) CHASER: a global chemical model of the troposphere 1. Model description. J Geophys Res 107(D21):4339. doi:10.1029/2001JD001113
Sudo K, Takahashi M, Akimoto H (2002b) CHASER: a global chemical model of the troposphere 2. Model results and evaluation. J Geophys Res 107(21):4586. doi:10.1029/2001JD001114
Treut HL, Li Z-X (1991) Sensitivity of an atmospheric general circulation model to prescribed SST changes: feedback effects associated with the simulation of cloud optical properties. Clim Dyn 5:175–187

Uno I, Carmichael G, Streets D et al (2003) Regional chemical weather forecasting system CFORS: Model descriptions and analysis of surface observations at Japanese island stations during the ACE-Asia experiment. J Geophys Res 108(D23):8668. doi:10.1029/2002JD002845

US EPA (1991) Guideline for regulatory application of the urban airshead model. Number EPA-450/4-91-013 in US EPA Report, Office of Air and Radiation, Office of Air Quality Planning and Standards, Technical Support Division, Research Triangle Park, North Carolina, US

Yurganov L et al (2005) Increased Northern Hemispheric carbon monoxide burden in the troposphere in 2002 and 2003 detected from the ground and from space. Atmos Chem Phys 5:563–573

Chapter 18
Operational Ozone Forecasts for Austria

Marcus Hirtl, K. Baumann-Stanzer, and B.C. Krüger

18.1 Introduction

Daily ozone forecasts for Austria have been run in an operational mode during summer 2005 and 2006. The model system has been set up for Austria in cooperation with the Central Institute for Meteorology and Geodynamic (ZAMG) and the University of Natural Resources and Applied Life Sciences (BOKU). The meteorological fields are supplied by the limited area model ALADIN-Austria (run twice a day at ZAMG). The 48-h forecasts are computed on domains with a horizontal resolution of 9.6 km (covering Austria) and 29 km (covering Central Europe). The dispersion modelling is done with the Comprehensive Air quality Model CAMx (version 4.20) with the SAPRC99-mechanism for gas phase chemistry.

18.2 Description of the Modelling System

The new air quality model system consists of three parts that are linked off-line together (Fig. 18.1). The combination of the two major parts, the meteorological input provided by ALADIN and the chemical model CAMx, was implemented for the first time in this study.

CAMx (Comprehensive Air quality Model with extensions, http://www.camx.com) simulates the emission, dispersion, chemical reaction, and removal of pollutants in the troposphere by solving the pollutant continuity equation for each chemical species on a system of nested 3D grids. A two grid nesting is used with a coarse grid over Europe and a finer grid for the core area covering Austria with the best possible spatial resolution of 9.6 km (according to the present grid of ALADIN-Austria).

M. Hirtl (✉)
Central Institute for Meteorology and Geodynamics (ZAMG), Hohe Warte 38, 1190, Vienna, Austria
e-mail: Marcus.Hirtl@zamg.ac.at

A. Baklanov et al. (eds.), *Integrated Systems of Meso-Meteorological and Chemical Transport Models*, DOI 10.1007/978-3-642-13980-2_18,
© Springer-Verlag Berlin Heidelberg 2011

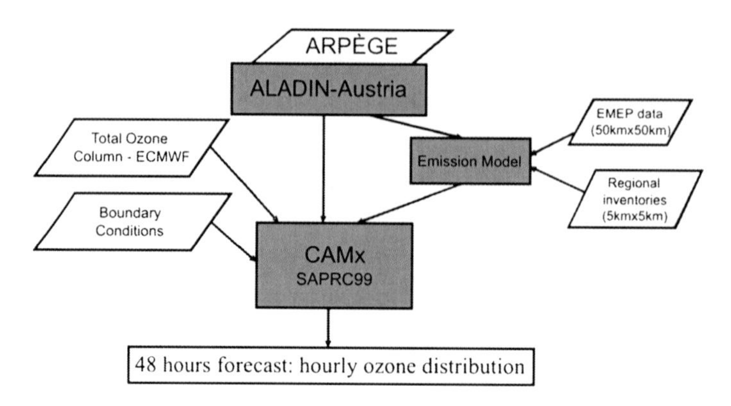

Fig. 18.1 Overview of the modelling system

The meteorological fields are supplied by the limited area model ALADIN-Austria (http://www.cnrm.meteo.fr/aladin/). It is run twice a day at the ZAMG and renders forecasts for 48 h. The meteorological fields have a temporal resolution of 1 h. The data is provided on 45 levels, and model has a horizontal resolution of 9.6 km. Fields of wind, temperature, pressure, convective and large scale precipitation, snow cover, solar radiation and specific humidity are extracted directly from the ALADIN dataset. The other fields, cloud optical depth, cloud water- and precipitation water content have to be parameterised (Seinfeld and Pandis 1998) from the ALADIN output.

The model system generally uses EMEP emissions. For the countries – Austria, Czech Republic, Slovakia and Hungary – the original 50 × 50 km data are downscaled to 5 × 5 km based on an inventory from 1995 (Winiwarter and Zueger 1996). The EMEP data for 1999 (used for 2005 forecast) have been substituted by data for 2003 (Vestreng et al. 2004). In addition, a new highly resolved emission inventory for the city of Vienna, Austria (Orthofer et al. 2005) is used for this area.

The boundary conditions of the coarse grid were estimated from the forecast of the previous day. This method is compared with constant boundary conditions using average summer values. Total ozone column data was obtained from ECMWF data.

18.3 Operational Forecasts 2006

High ozone values are most frequently observed in the eastern parts of Austria, where warnings for values above the information or the alarm threshold are launched for ozone region 1 (covering lower Austria, Vienna and Burgenland).

Figure 18.2 shows the predicted maximum concentrations for ozone region 1. The course of ozone concentration from 1 day (Tag 1) and 2-days (Tag 2) model forecasts as well as the results from a backup run which considers constant boundary conditions only are compared to measurements (at 43 air quality stations).

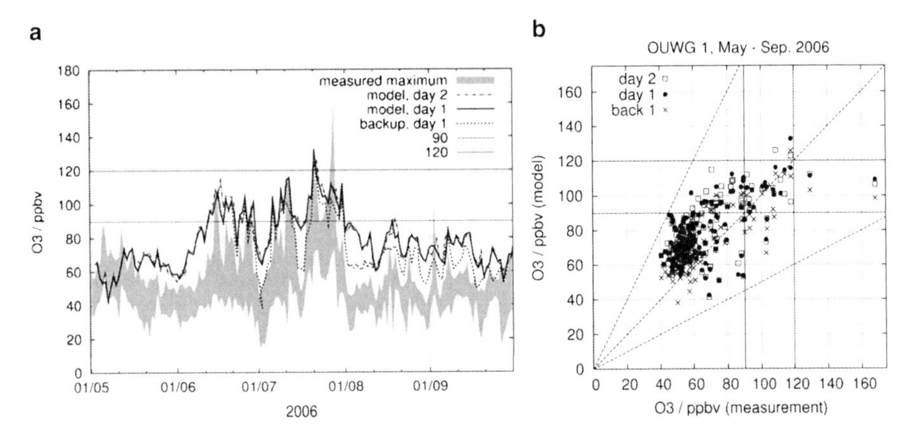

Fig. 18.2 (**a**) *Gray area*: range between highest and lowest maximum observations (hourly average) at stations within the region; with maximum predicted on the same and previous days; backup run (constant boundary conditions). (**b**): Scatter-diagram of daily ozone maxima predicted versus observed in ozone region 1 for 2006 (maximum predicted on previous and on the same day, as well as maximum predicted by backup run; lines: information and alert threshold (*Directive 2002/3/EC*))

Exceedances of threshold of 180 $\mu g/m^3$ ($\approx$90 ppbv) occurred in 2006 between middle of June and end of July. After that period the predicted concentrations were higher than the measurements. The light-blue line shows that the model run considering constant boundary conditions performs slightly better than the dynamic approach during that period.

According to the station measurements, the information threshold has been exceeded on 17 days in the period depicted in Fig. 18.2a. Note that 16 of these exceedances were forecasted correctly. The alert exceedance predicted for July 21 was not observed but the measured values were just below the threshold value. The daily maxima above the alert threshold which were observed on 27–28 July 2006, were not predicted by the model. On these days, local peak emissions obviously caused a sudden and local increase of ozone concentrations around noon for 1–2 h.

Figure 18.2b shows a scatter diagram of observed vs. modelled ozone in region 1. Depicted are daily maxima for the two forecast days and the backup run. Information (90 ppbv) and alarm threshold (120 ppbv) are marked by magenta lines (horizontal and vertical). Most of the values lie above the 1:1 line which means that the model tends to predict higher concentrations than observed. Table 18.1 summarizes the metrics of some selected stations in ozone region 1. The correlation and the standard deviation between observations and model prediction of the two forecast days as well as the backup run are depicted. The last line shows the values for all 43 stations in ozone region 1.

The correlation between the two forecast days with the measurements is practically the same. It seems that the backup run correlates better with the observations for the selected stations as well as for the whole region studied. The standard deviation is also smaller for backup run.

Table 18.1 Correlation (r) and standard deviation (s) of selected stations in ozone region 1

	Forecast day 1		Forecast day 2		Backup day 1	
	r	s	r	s	r	s
Eisenstadt	0.62	12.16	0.61	12.54	0.73	10.26
Kittsee	0.63	13.47	0.64	13.15	0.72	11.75
Himberg	0.69	13.10	0.69	13.29	0.76	11.51
Schwechat	0.71	13.85	0.71	13.99	0.78	12.10
Tulln	0.65	14.27	0.65	13.84	0.70	12.49
Wiener Neustadt	0.54	12.23	0.55	12.27	0.64	10.48
Hohe Warte	0.68	13.33	0.68	13.53	0.75	11.56
Lobau	0.71	13.71	0.71	13.99	0.78	12.03
Average	0.59	13.03	0.59	13.00	0.70	10.93

The statistical comparisons between ozone forecasts and measurements for all stations in eastern Austria show correlation coefficients between 0.4 and 0.7 and standard errors around 12 ppbv for 2006.

It was investigated how accurate exceedances of the information threshold could be predicted by the model. The following combinations were considered, i.e. days when:

- The highest value of the observations as well as the model forecasts are bellow the information threshold (90 ppbv) in ozone region 1
- Observation and model lie above the threshold
- The observations lie below the threshold, the model above it
- The observations lie above the threshold, the model below it

Table 18.2 shows the values for the respective months. Operational forecasts for 2006 and 2005 are compared with results of the backup run.

Although the predicted ozone values tend to be higher than the observations in 2006 the hit rate to predict the exceedance of the information threshold is 88%. This is slightly better than the backup run in 2006 (84.8%) and it is worse than in 2005 (90.9%). These values include also months with low ozone values in Austria. During May, August, and September, no exceedances of the information threshold occurred due to meteorological conditions. Considering only June and July 2006 (in total 61 day) the hit rate would be 70.5 and 63.6% for the operational and backup runs, respectively.

18.4 Conclusion

Daily ozone forecasts for Austria have been run during summer 2005 and 2006. The results of the forecasts have been evaluated with measurements of the Austrian air quality network for eastern parts of Austria. Generally the observed exceedances of the information threshold are re-produced well by the model. Days with exceedances of the information threshold were predicted by the model with a probability of 88% during the summer period in 2006. The daily maxima above the alert

Table 18.2 Hit-rate for the exceedance of the information threshold

Obs.	Mod.	May	June	July	Aug.	Sep.	Total	%
Forecast 2006, day 1								
<90	<90	28	16	11	31	30	116	88.0
>90	>90	0	3	13	0	0	16	
<90	>90	0	11	6	0	0	17	12.0
>90	<90	0	0	1	0	0	1	
Total		28	30	31	31	30	150	
Backup-run 2006, day 1								
<90	<90	–	7	10	31	30	78	84.8
>90	>90	–	1	10	0	0	11	
<90	>90	–	4	7	0	0	11	15.2
>90	<90	–	1	4	0	0	5	
Total		–	13	31	31	30	105	
Forecast 2005, day 1								
<90	<90	–	3	25	30	27	84	90.9
>90	>90	–	2	4	0	0	6	
<90	>90	–	2	0	1	2	6	9.1
>90	<90	–	0	2	0	1	3	
Total		–	7	31	31	30	99	

threshold were not predicted by the model probably due to rather low resolution. The ozone forecasts are continued with an improved model system in the ozone season 2007. The emission inventories as well as the CAMx model will be updated. Additionally a new approach to obtain the boundary conditions from climatological average values will be applied and test runs with increased spatial resolution will be conducted.

Acknowledgements This work has been funded by the Bundesministerium für Land- und Forstwirtschaft; Umwelt und Wasserwirtschaft (BMLFUW); the Magistrat der Stadt Wien; the Amts der Niederösterreichischen Landesregierung, Burgenländischen Landesregierung, Steiermärkischen Landeregierung, Kärntner Landesregierung, Oberösterreichischen Landesregierung, and Salzburger Landesregierung.

References

Orthofer R, Humer H, Winiwarter W, Kutschera P, Loibl W, Strasser T, Peters-Anders J (2005) emikat.at – Emissionsdatenmanagement für die Stadt Wien. ARC system research, Bericht ARC-sys-0049, Seibersdorf, Austria

Seinfeld JH, Pandis SN (1998) Atmospheric chemistry and physics, from air pollution to climate change (1173–1174). Wiley, New York

Vestreng V, Adams M, Goodwin J (2004) Inventory Review 2004, Emission Data reported to CLRTAP and under the NEC Directive, EMEP/EEA Joint Review Report, EMEP/MSC-W Note 1/2004. ISSN 0804-2446

Winiwarter W, Zueger J (1996) Pannonisches Ozonprojekt, Teilprojekt Emissionen. Endbericht. Report OEFZS-A-3817, Austrian Research Center, Seibersdorf

Chapter 19
Impact of Nesting Methods on Model Performance

Ursula Bungert and K. Heinke Schlünzen

19.1 Introduction

The air quality directive of the European commission demands maps on concentrations and exceedances in different detail. For this purpose numerical models can be used. Some model systems are already adjusted to deliver the corresponding maps. For instance, the model system M-SYS consists of three mesoscale and one microscale model areas and applies one-way-nesting for meteorology and chemistry (Trukenmüller et al. 2004).

To calculate concentration data, reliable meteorological fields are needed, which should be calculated with the same resolution as the concentration maps. Lenz et al. (2000) have shown that the concentration fields are more sensitive to the description of the meteorological fields at the lateral boundaries than to the concentration fluxes over these boundaries. Nesting the meteorological fields, corresponding to the use of meteorological model results received on a coarser grid as time-dependent boundary values, needs to prescribe the boundary values as realistic as possible. In this work, we consider only the nesting of the meteorological fields and the influence of different update intervals of the forcing data on the model performance of nested simulations.

19.2 Method

The multiscale meteorology and chemistry model system M-SYS consists of the mesoscale models: MEsoscale TRAnsport and Stream (METRAS; Schlünzen 1990; Schlünzen and Katzfey 2003) and MEsoscale Chemistry Transport Model (MECTM; Müller et al. 2000; Schlünzen and Meyer 2007), which are used in

K. Heinke Schlünzen (✉)
Meteorological Institute, KlimaCampus, University of Hamburg, Bundesstr. 55, 20146 Hamburg, Germany
e-mail: heinke.schluenzen@zmaw.de

A. Baklanov et al. (eds.), *Integrated Systems of Meso-Meteorological and Chemical Transport Models*, DOI 10.1007/978-3-642-13980-2_19,
© Springer-Verlag Berlin Heidelberg 2011

different resolutions, and the obstacle-resolving microscale models MITRAS and MICTM (Schlünzen et al. 2003; López et al. 2005). Meteorology and chemistry transport are coupled off-line. Because the same grid, the same model physics and parameterizations are used, no interfaces are needed for the meteorology-chemistry coupling. Model results are, beside others, maps of concentration fields on different scales with different resolutions.

Although, the nesting is applied in M-SYS at every time step, the forcing fields are only available at time intervals that are much longer than the time step used in the model. During these intervals, the forcing fields are linearly interpolated. To investigate if the linear interpolation represents the development of the atmospheric fields in a realistic way, the mesoscale atmospheric model METRAS is used in different resolutions. The results of coarser grid simulations are used as forcing fields for the nested simulations in higher resolution. The time interval for writing the coarser model results determines the duration in which the forcing fields in the higher resolution simulations can be updated. By using the same model for calculating forcing data and performing the nested simulations on a higher resolving grid it is possible to perform well-controlled sensitivity studies on the effect of update interval on high resolution model results.

For nesting of meteorological models and for coupling of meteorology and chemistry transport it is important to know, how often the atmospheric forcing fields should be updated to sufficiently represent the non-linear processes triggered from the boundaries. If the update interval for the forcing fields is very small, the non-linear processes in the atmosphere should be well represented during these short intervals where a linear interpolation is applied. By continuous updating we eventually receive a model with multiple grids, which is a valid (but expensive) approach. However, despite costs this approach is not always possible, since limited area models need at some point lateral boundary values. At the outermost domain the forcing data are only available at specific time intervals. It needs to be known in which frequency these forcing data should be available. In addition, reading of forcing data needs additional time on the computer and thus, should be reduced to the necessary amount. Data update should be performed as much as necessary and as little as possible.

Changes in the atmospheric fields happen on different time scales and with different speed. Therefore, we adapt the time intervals for writing the model results (that are the forcing data for nested simulations) to the time scales, in which the atmospheric fields change, instead of using short but constant time intervals. If significant changes happen on short time scales, the results should be written more often than for more or less steady conditions.

The horizontal wind components, potential temperature and specific humidity are used in METRAS as forcing fields. Changes in the scalar quantities temperature and humidity are mainly induced by advection and diffusion, i.e. processes that depend on the wind. Therefore, we define the conditions for writing the model results only in dependence of changes in the horizontal wind components. The model results were alternatively written at regular intervals (3 h, 6 h), if the accelerations (changes in velocity) in 80 and 20% of the grid points are less than 5×10^{-5} and 5×10^{-6} m/s^2, respectively

19.3 Simulation Set-up

The outlined method of a situation-controlled writing of the model results is applied to a period in August 2003. During the simulated period (29–31 August 2003), a trough was extended from Spain to the Arctic northeast of Finland, lying between high-pressure systems over the Atlantic Ocean and south-eastern, and then later Eastern Europe. In between this trough, a small scale low has developed originally lying over Belgium on 28th August, and then passed in west-east direction over northern Germany. Caused by this low, a significant precipitation occurred in this area.

The model areas used for simulating this case are shown in Fig. 19.1. The large area (Fig. 19.1a) covering significant parts of Europe has a horizontal resolution of 18 km, while a grid size in the small area (Fig. 19.1b) is 6 km. Model simulations are started for 20 CET, 28 August 2003. Comparisons are performed starting 1 h after simulation begins (from 21 CET, 28 August 2003).

Several simulations on the 18 km grid that use different conditions for writing the model output produce the forcing data for the nested simulations that use a 6 km resolution. The 6 km simulations are summarized in Table 19.1. As seen, besides the four simulations that are nested in METRAS 18 km results, two additional simulations were also performed. In the simulation 5 (eu_6km_ana) the forcing fields are derived from analyses that are available every 12 h; thus, a very large update interval is used. This update interval cannot be changed. Simulation 6 (eu_6km_nonesting) uses no nesting and no heterogeneous initialisation, but only

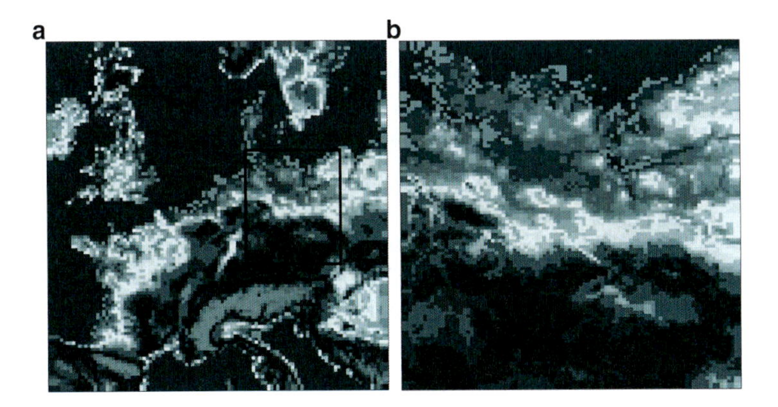

Fig. 19.1 The model areas used for the coarse grid run yielding the forcing data (**a**) and for the nested simulations for the case "low pressure system over Europe" (**b**). The positioning of the nested area is shown by the frame in (**a**)

Table 19.1 Simulations performed with the high-resolution model area shown in Fig. 19.1b

	Simulation name	Forcing data from	Update interval
1	eu_6km_3h	Coarse grid METRAS run	3 h
2	eu_6km_6h	Coarse grid METRAS run	6 h
3	eu_6km_lcout80	Coarse grid METRAS run	Depending on results (80%)
4	eu_6km_lcout20	Coarse grid METRAS run	Depending on results (20%)
5	eu_6km_ana	Analysis	12 h
6	eu_6km_nonesting	–	–

integrates the initial profile forward in time. This is a sensitivity study to allow evaluating if the large-scale situation has any impact on the high-resolution results.

19.4 Simulation Results

The results of the all simulations given in Table 19.1 are compared to each other and to a reference case which is a simulation with high resolution in the whole domain (Fig. 19.1a). The hit rates (in %) for wind speed and direction, and temperature were calculated for every hour of the simulation using allowed deviations. The desired accuracies DA used for calculating the hit rate (from Cox et al. 1998) are: ±2°C for the air and dew point temperatures, ±30° for the wind direction, ±1.7 hPa for the pressure, and ±1 and ±2.5 m/s for the wind speeds of less than and more than 10 m/s, respectively. All hit rates are based on comparisons with the reference case and include all grid points (about 3×10^6) of the model domain in Fig. 19.1b.

Values for the hit rates are shown in Fig. 19.2 for the six different simulations. The simulation 6 has by far the largest deviations from the reference case. This shows that the result is indeed sensitive to changes in the large-scale situation. In addition, also simulation 5 yields quite low hit rates. There are two probably reasons for this: (a) the update interval of the forcing data is too long compared to the changes in the large-scale situation; (b) the forcing data derived from the analysis differ from the coarse grid simulation results and therefore lead to the low hit rates which are based on the results of the reference simulation.

The simulations nested in the coarse grid METRAS results agree in a similar way with the reference simulation in the second half of the simulation time (after about 21 h; i.e. 18 CET of 29 August 2003). At this time the performance of simulation 5 is somewhat closer to the other nested simulations than before. This might be a hint that the nesting becomes less relevant and the situation is more locally driven. In the first 21 h of the simulation the two simulations with constant update intervals (3 h, 6 h) are closest to the reference case, while the adaptive update simulations (3, 4) show a high variability in performance. This is a hint that the acceleration is probably not a reliable measure to determine update intervals. The best performance is received in the present case study for a nesting every 3 h.

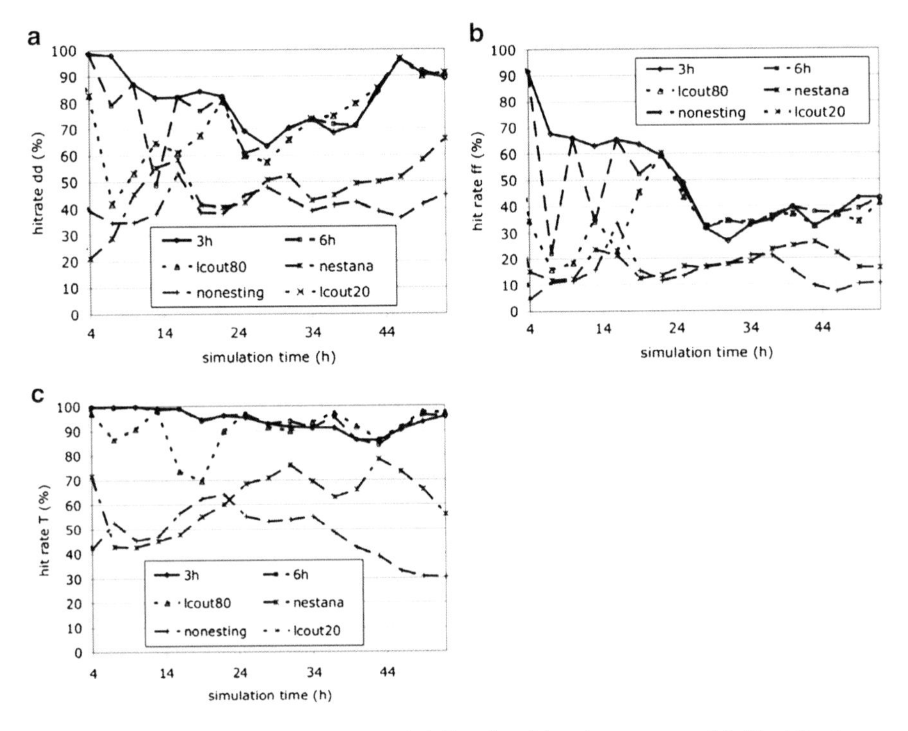

Fig. 19.2 Hit rates for wind speed (**a**), wind direction (**b**) and temperature (**c**). The hit rates are calculated by comparing results of the simulations with the high-resolution reference case

19.5 Conclusions and Outlook

METRAS is able to yield simulation results on different scales and details. Nesting is a helpful tool to improve the model performance if the forcing data represent the meteorological situation in a realistic way. The current studies on the impact of update interval lengths on model performance lead to the following conclusions:

- In general, shorter update intervals lead to higher model performance
- Consecutive, but very different intervals seem to reduce the positive effect
- With increasing forecast time, the uncertainty in the forecast (that is resulting from the update of the boundary values) seems to decrease

Especially the last result is surprising, since it is generally assumed that for longer forecast with limited area models the boundary values become more relevant. Therefore, it needs to be evaluated, if this is a result only true for the situation studied or if it is more general.

A next step to further investigate the influence of different update intervals will be to define a characteristic time for the occurring changes and using this characteristic time to control the writing of the data that are used as forcing fields in the high resolution simulations. First results seem to be promising in enhancing model performance compared to the constant update interval of 3 h.

Acknowledgments This work has been supported by the German Research Foundation (DFG) in the framework of the project SCHL499-2 and the Sonderforschungsbereich 512 "Cyclones and the North Atlantic Climate System". This work is a contribution to COST-728 Action.

References

Cox R, Bauer BL, Smith T (1998) Mesoscale model intercomparison. Bull Am Meteorol Soc 79:265–283

Lenz C-J, Müller F, Schlünzen KH (2000) The sensitivity of mesoscale chemistry transport model results to boundary values. Environ Monit Assess 65:287–298

López SD, Lüpkes C, Schlünzen KH (2005) The effects of different k-e-closures on the results of a micro-scale model for the flow in the obstacle layer. Meteorol Z 14:839–848

Müller F, Schlünzen KH, Schatzmann M (2000) Test of numerical solvers for chemical reaction mechanisms in 3D air quality models. Environ Model Softw 15:639–646

Schlünzen KH (1990) Numerical studies on the inland penetration of sea breeze fronts at a coastline with tidally flooded mudflats. Beitr Phys Atmos 63:243–256

Schlünzen KH, Katzfey JJ (2003) Relevance of subgrid-scale land-use effects for mesoscale models. Tellus 55A:232–246

Schlünzen KH, Meyer EMI (2007) Impacts of meteorological situations and chemical reactions on daily dry deposition of nitrogen into the Southern North Sea. Atmos Environ 41–2:289–302

Schlünzen KH, Hinneburg D, Knoth O, Lambrecht M, Leitl B, Lopez S, Lüpkes C, Panskus H, Renner E, Schatzmann M, Schoenemeyer T, Trepte S, Wolke R (2003) Flow and transport in the obstacle layer – first results of the microscale model MITRAS. J Atmos Chem 44:113–130

Trukenmüller A, Grawe D, Schlünzen KH (2004) A model system for the assessment of ambient air quality conforming to EC directives. Meteorol Z 13(5):387–394

Chapter 20
Running the SILAM Model Comparatively with ECMWF and HIRLAM Meteorological Fields: A Case Study in Lapland

Marko Kaasik, M. Prank, and M. Sofiev

20.1 Introduction

This case study is intended to clarify the range of uncertainties in meso-scale atmospheric dispersion modelling due to different dispersion schemes (i.e. Lagrangian and Eulerian) and meteorological input. Propagation of a plume from a single point source was recorded in a monitoring station with a high temporal resolution. Thus, this case gives a deeper insight into variability of reproduction of instantaneous advection and dispersion conditions without diluting impact of numerous sources and time averaging.

20.2 Methods

This modelling study is based on the aerosol measurement campaign carried out during April–May 2003 at the Värriö monitoring station (67°46′N, 29°35′E; Finland, Eastern Lapland) located close to the Russian border (Ruuskanen et al. 2007). On the generally low aerosol background of Arctic spring only a few pollution episodes were observed (Fig. 20.1). In this paper we will focus on the highest episodes (i.e. up to 30 $\mu g/m^3$ of PM10 on May 2–3).

First, the SILAM model (Sofiev et al. 2006) was applied in adjoint mode to identify the potential sources of pollution. It was found that the aerosol peak of May 2–3 most probably originated from the Nikel metallurgy factory (Kola Peninsula, Russia) located about 200 km north from Värriö (Kaasik et al. 2007). Then the SILAM model was applied in a forward mode comparatively with the ECMWF and HIRLAM (FMI) meteorological datasets: EMEP emission data on sulphate and PM, sea salt emissions calculated by SILAM, emission model based on

M. Kaasik (✉)
Faculty of Science and Technology, Institute of Physics, University of Tartu, Tähe 4, 51014 Tartu, Estonia
e-mail: marko.kaasik@ut.ee

A. Baklanov et al. (eds.), *Integrated Systems of Meso-Meteorological and Chemical Transport Models*, DOI 10.1007/978-3-642-13980-2_20,
© Springer-Verlag Berlin Heidelberg 2011

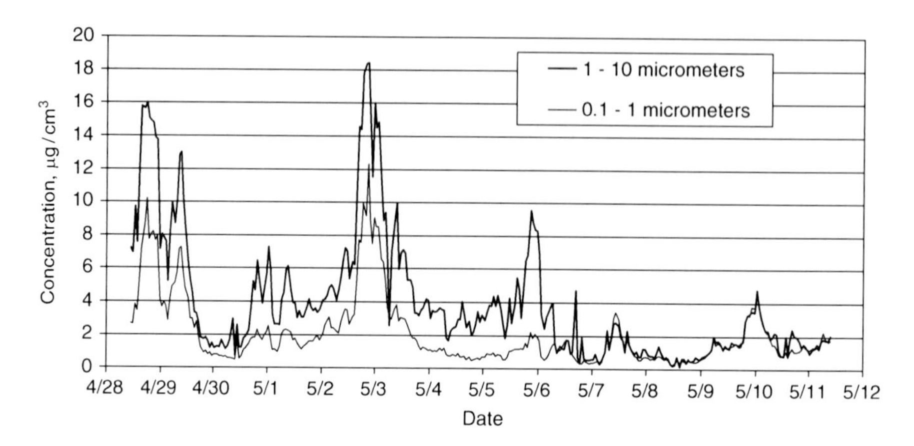

Fig. 20.1 Measured surface-level concentrations of PM10 and PM1 in Värriö during April–May 2003

(Mårtensson et al. 2003). The well-tested Lagrangian kernel (Lagrangian particle model, version 3.7) and new Eulerian kernel (test version 4.1 released in 2007, presently under evaluation) of SILAM were applied with both meteorological datasets; thus, producing ensemble of four runs. Parameters of a highly buoyant plume from high stack were assigned: source height distributed from 200 to 1,000 m.

20.3 Results

In the large scale the concentration patterns produced in all four runs were rather similar, but some differences concerning the May 2–3 peak appeared critical for local measurement-modelling comparison (see Fig. 20.2):

- The Lagrangian plume with ECMWF data narrowly missed the monitoring station passing eastwards; but with HIRLAM data plume matched the monitoring station, predicted concentrations and timing rather similar to measured ones
- The Eulerian plume with ECMWF data matched the monitoring station with slight delay, predicted concentrations were overestimated; but with HIRLAM data plume matched the monitoring station with a slight delay, predicted concentrations were highly overestimated

It appeared that contribution of other emissions besides sulphates to this particular event was negligible. Thus, only the measured aerosol below 1 μm size limit (typical for sulphate aerosol) was taken for comparison.

Surface-level concentration maps of sulphate computed with (Lagrangian) SILAM 3.7 are presented in Fig. 20.3. The ECMWF data run gives a narrow plume directly to south from the plant, which misses the Värriö site just by one grid cell (20 km). The HIRLAM data run forces the plume to travel, at first, to south, and then to south-west passing over Värriö.

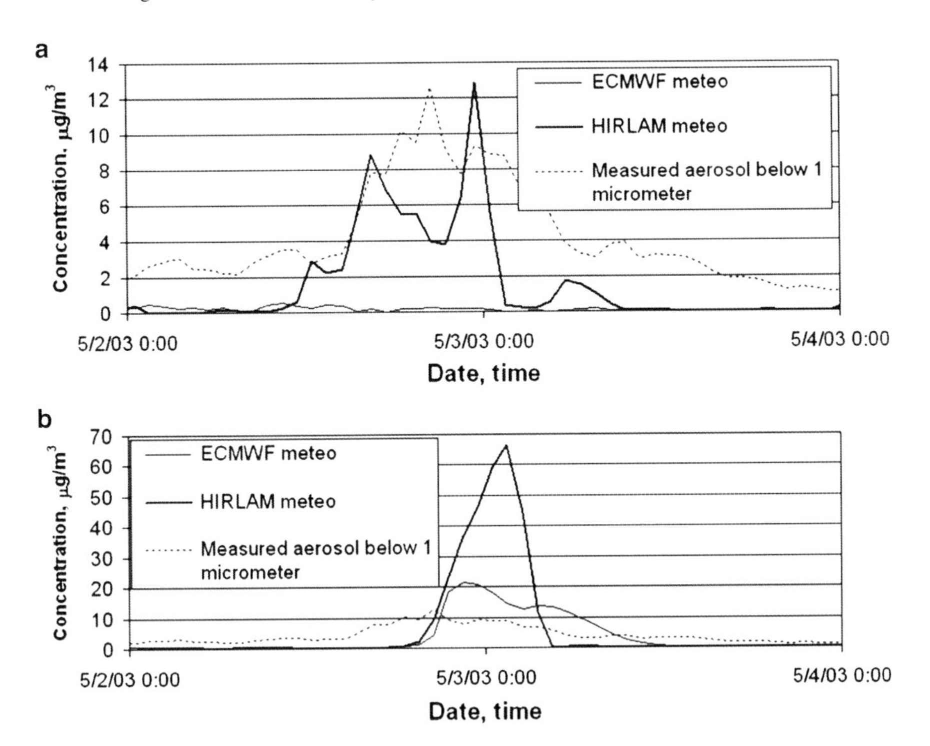

Fig. 20.2 Surface-level concentrations of sulphate in Värriö during the peak event, modeled with SILAM: (**a**) Lagrangian kernel, (**b**) Eulerian kernel with ECMWF and HIRLAM meteorological input, compared to the measured concentrations of aerosol particles with diameter below 1 m

In general, the Eulerian scheme (version 4) produced much wider horizontal spread than the Lagrangian (Fig. 20.4), despite the higher concentration peaks. Runs with both meteorological drivers showed the sulphate polluted plume over Värriö, but with about 5-h delay compared to both measurements and Lagrangian (HIRLAM data) run.

Differences were found in vertical spread as well. In the Lagrangian run (Fig. 20.5) the concentration fields in the lower free troposphere followed the surface-level concentrations in general. This reflected the simplified vertical structure of this kernel, in particular, well-mixing assumption for the boundary layer and fixed diffusion term in the free troposphere. In the Eulerian run (Fig. 20.6), the higher-level modeled concentrations were patchy and less correlated with the near-surface fields in comparison with the Lagrangian run.

20.4 Discussion

The weather situation during the peak was complicated due to a high-pressure system with a centre located close to the monitoring site. Thus, wind was weak, changing rapidly in space and time. In such conditions small discrepancies between

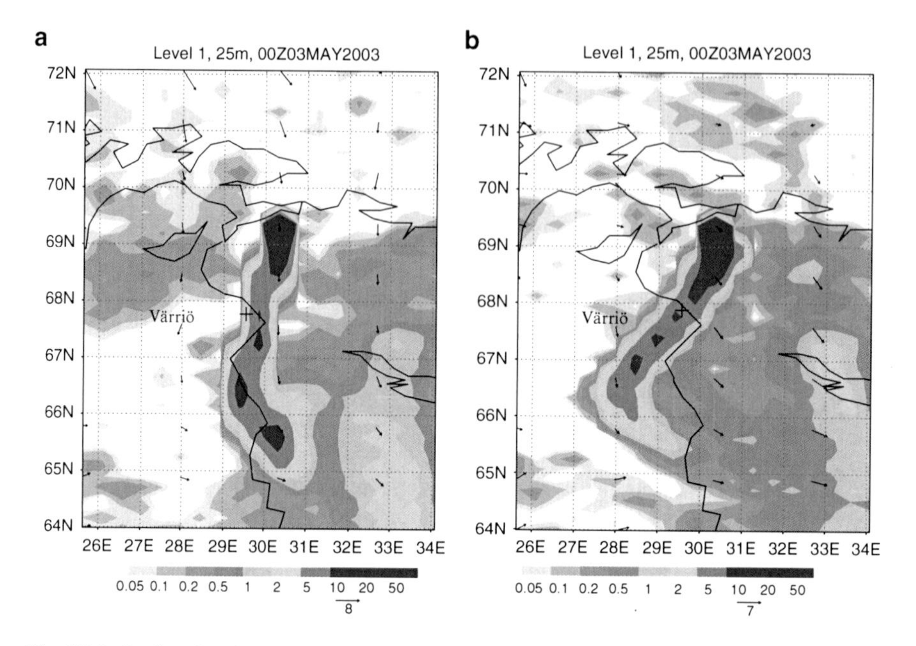

Fig. 20.3 Surface-level concentration fields of sulphate, 00:00 at May 3, 2003, calculated with Lagrangian SILAM 3.7: (**a**) With ECMWF meteorological fields, (**b**) With HIRLAM (FMI) meteorological fields. *Arrows* – wind at 25 m

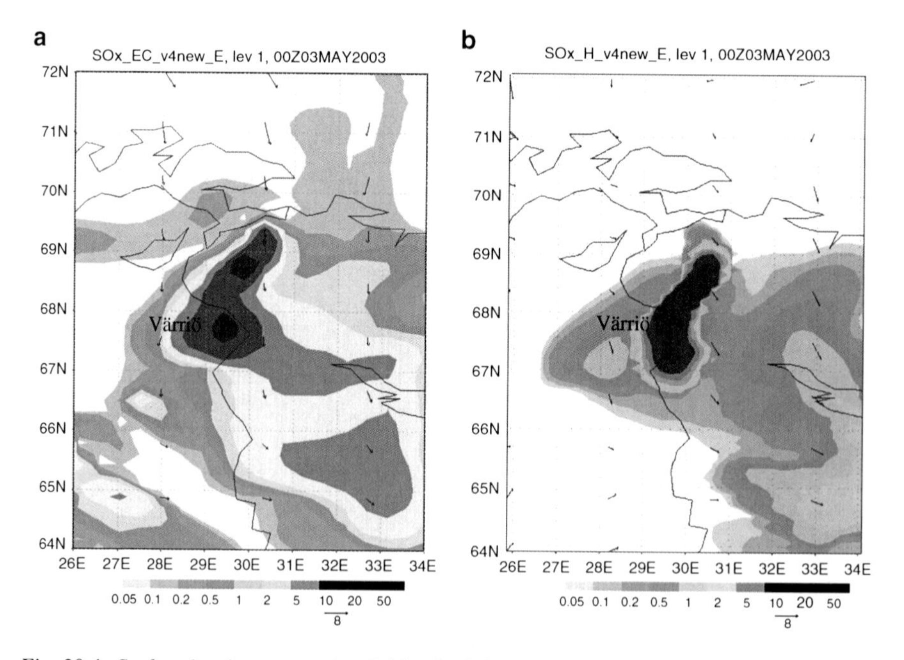

Fig. 20.4 Surface-level concentration fields of sulphate, 00:00 at May 3, 2003, calculated with Eulerian SILAM 4: (**a**) With ECMWF meteorological fields, (**b**) With HIRLAM (FMI) meteorological fields. *Arrows* – wind at 25 m

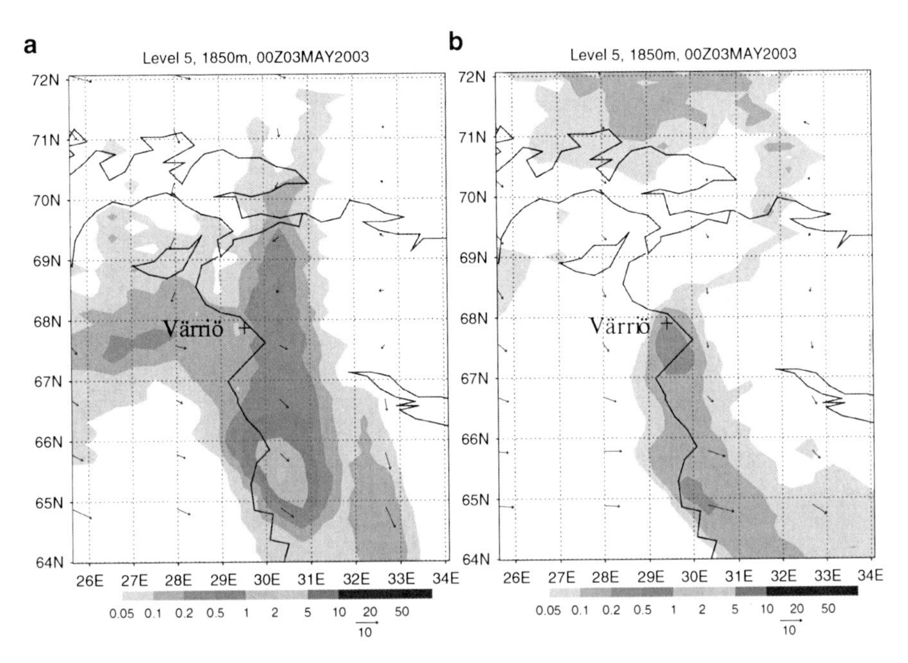

Fig. 20.5 1,850 m fields of sulphate, 00:00 at May 3, 2003, calculated with Lagrangian SILAM 3.7: (**a**) With ECMWF meteorological fields, (**b**) With HIRLAM (FMI) meteorological fields. *Arrows* – wind at 1,850 m

the meteorological fields and ways to treat these in the dispersion model become essential.

In this case study the Lagrangian run with HIRLAM data appears to match the observations better than the others. However, its plume was too narrow and its dispersion velocity seems to be somewhat over-predicted (Fig. 20.2a). Also, it is obvious that the Lagrangian run showed higher variability of the concentrations than the Eulerian one. The Eulerian plume appears much wider, traveling slower due to larger mass fraction near the surface and, thus, matching the measurement site more certainly. It also resulted in a rather smoother shape of concentration time series (Fig. 20.2b).

The Eulerian run with ECMWF meteodata appears closer to the shape of observed time series while the stronger vertical motions predicted by the HIRLAM system caused a strong peak in the concentrations.

20.5 Conclusions

The current exercise highlights the objective difficulties in predicting the short-term episodes originated from a single nearly-point source and observed at a single site. It is, however, evident that an ensemble of four partly independent chemical-weather

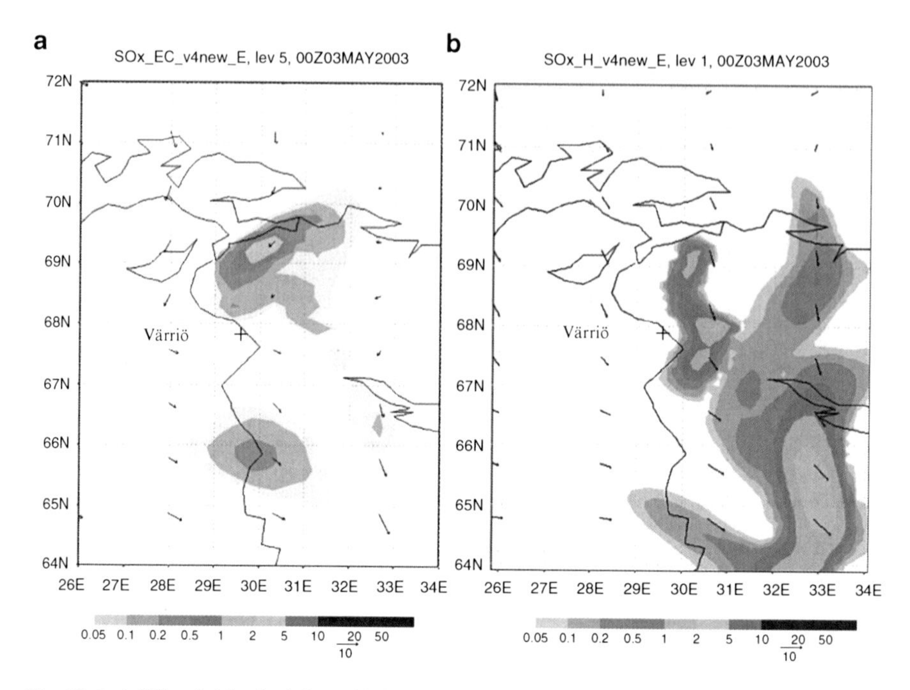

Fig. 20.6 1,850 m fields of sulphate, 00:00 at May 3, 2003, calculated with Eulerian SILAM 4: (**a**) With ECMWF meteorological fields, (**b**) With HIRLAM (FMI) meteorological fields. *Arrows* – wind at 1,850 m

predictions appeared able to show high probability of the episode, its stochastic features, and a potential range of uncertainties in the results of simulations.

Acknowledgements This study was supported by the Estonian Science Foundation, Grant 7005 and performed within the scope of the SILAM system evaluation for EU-GEMS and ESA-PROMOTE projects.

References

Kaasik M, Prank M, Kukkonen J, Sofiev M (2007) A suggested correction to the EMEP database, regarding the location of a major industrial air pollution source in Kola Peninsula. In: Miranda AI, Borrego C (Toim.) 29th NATO/SPS international technical meeting on air pollution modelling and its application (Preprints), University of Aveiro, Aveiro, Portugal, 24–28 Sep, pp 333–339

Mårtensson EM, Nilsson ED, de Leeuw G, Cochen LH, Hansson H-C (2003) Laboratory simulations of the primary marine aerosol production. J Geophys Res 108(9):4297. doi:10.1029/2002JD002263

Ruuskanen TM, Kaasik M, Aalto PP, Hõrrak U, Vana M, Mårtensson M, Yoon Y, Keronen P, Mordas G, Nilsson D, O'Dowd C, Noppel M, Alliksaar T, Ivask J, Sofiev M, Prank M, Kulmala M (2007) Concentrations and fluxes of aerosol particles during LAPBIAT measurement campaign in Värriö field station. Atmos Chem Phys 7:3683–3700

Sofiev M, Siljamo P, Valkama I, Ilvonen M, Kukkonen J (2006) A dispersion modelling system SILAM and its evaluation against ETEX data. Atmos Environ 40:674–685

Part IV
Strategy for ACT-NWP Integrated Modeling

Chapter 21
HIRLAM/HARMONIE-Atmospheric Chemical Transport Models Integration

Alexander Baklanov, Sander Tijm, and Laura Rontu

The 'HIRLAM/HARMONIE-ACTMs Integration' discussion session was held on Tuesday (22 May 2007) during the COST-728-NetFAM 'Model Integration' workshop. Additionally to two overall introductory presentations by Alexander Baklanov, DMI (see at http://netfam.fmi.fi/Integ07/baklanov_prese2.pdf) and Sander Tijm, KNMI (see at http://netfam.fmi.fi/Integ07/tijm_prese.pdf) and round table discussions the following six specific presentations of the HIRLAM partners models were given (see the corresponding papers published in this volume):

- Enviro-HIRLAM Status and Evaluation of Differences between On-line and Off-line Models: Korsholm et al.
- Coupling of Air Quality and Weather Forecasting – Progress and Plans at met. no: Ødegaard et al.
- A Note on Using the Non-Hydrostatic Model AROME as a Driver for the MATCH Model: Robertson and Foltescu
- Overview of DMI ACT-NWP Modelling Systems: Baklanov et al.
- Aerosol Species in the AQ Forecasting System of FMI: Possibilities for Coupling with NWP Models: Sofiev et al.
- Running the SILAM Model Comparatively with ECMWF and HIRLAM Meteorological Fields – a Case Study in Lapland: Kaasik et al.

21.1 Introduction to HIRLAM-ACTM Integration

The new strategy for development of a new generation integrated Meteorology (MetM) and Atmospheric Chemical Transport Model (ACTM) systems for predicting atmospheric composition, meteorology and climate change has become more

A. Baklanov (✉)
Danish Meteorological Institute (DMI), Lyngbyvej 100, DK-2100 Copenhagen, Denmark
e-mail: alb@dmi.dk

A. Baklanov et al. (eds.), *Integrated Systems of Meso-Meteorological and Chemical Transport Models*, DOI 10.1007/978-3-642-13980-2_21,
© Springer-Verlag Berlin Heidelberg 2011

and more important in recent years (Grell et al. 2005; Jacobson, 2006; Baklanov and Korsholm, 2007). This combination is reasonable due to the facts that: (1) meteorology is an important source of uncertainty in air pollution and emergency preparedness modelling, (2) there are complex and combined effects of meteorological and pollution components on human health, (3) pollutants have effects, especially aerosols, on climate forcing and meteorological phenomena (radiation, clouds, precipitation, thunderstorms, etc.). So, this way of integrating modelling can be beneficial for model improvements in both communities: NWP and atmospheric environment forecasting.

The formulation of this strategy on the HIRLAM community level was the focus of a special session on during the COST-728/NetFAM workshop on "Integrated systems of meso-meteorological and chemical transport models" (Tijm et al. 2007). The main purpose of the section was to gather together the HIRLAM NWP and ACTM modelers, to discuss and build a joint strategy for developing integrated system(s) based on HIRLAM/HARMONIE. Although the HIRLAM consortium and ACTM modellers in the HIRLAM-organisations have some interest and initiatives in such an integration, they work separately and have a very low level of coordination and cooperation in this area.

There are several attempts in this direction, including the following

- DMI is actively working with development of the fully on-line integrated system Enviro-HIRLAM, that considers chemistry, aerosol forcing mechanisms, etc.
- Most of HIRLAM-member institutes are using national HIRLAM NWP outputs as meteo-drivers for their ACTP modelling and air quality forecasting activities. They have already attempted to build off-line integrations of HIRLAM with their own ACTMs (CAC, Chimere, DERMA, EMEP, MATCH, SILAM).
- Such a work was also included in the HIRLAM-A development plan (S4.10/4.5 Task: Coupling with atmospheric chemistry).
- The 'Integration' WG2 in COST 728 involves 5 HIRLAM-member institute representatives (DMI, FMI, Met.no, SMHI, Estonian Tartu Univ.), and they are working to consolidate and coordinate joint efforts for coupling HIRLAM with ACTMs.
- Nordic Network on Fine-scale Atmospheric Modelling (NetFAM), involving 20 Nordic, Baltic and French meteorological institutes from universities and weather services, organised a summer school and workshop on "Integrated Modelling of Meteorological and Chemical Transport Processes / Impact of Chemical Weather on Numerical Weather Prediction and Climate Modelling" in July 2008 in Zelenogorsk, near St. Petersburg, Russia (http://netfam.fmi.fi/YSSS08).

The above approach can lead to improvements in the use of HIRLAM in ACTM as well as ACTM results utilization in HIRLAM for NWP refinement. At least two different tasks for HIRLAM/HARMONIE-ACTM cooperation are important to consider:

1. Improvement of HIRLAM outputs (as well as PBL schemes there) for ACT modelling applications and correspondingly improvement of ACT models (for different offline ACT models, like CAC, DACFOS, DERMA, EMEP, MATCH, SILAM)
2. Improvement of NWP itself by implementation of ACTMs and aerosol/gases forcing/feedback mechanisms into HIRLAM (mostly by online integration, like in Enviro-HIRLAM) as well as by further development of the radiation and cloud microphysics modules to introduce feedback mechanisms.

21.2 On-Line and Off-Line Coupling of HIRLAM and ACTM

The integration/coupling of the NWP HIRLAM/HARMONIE and ACTM models can be realized by different ways using the on-line and off-line modelling approaches (in more details the approaches and strategy are discussed in Baklanov 2008).

One-way integration (off-line):

- HIRLAM meteo-fields as a driver for CTM (this way is already used by many air pollution modelers in the HIRLAM countries)
- ACTM chemical composition fields as a driver for regional climate modelling (e.g. for aerosol forcing on meteo-processes, it could also be realized for NWP, e.g. for HIRLAM)

Two-way integration:

- Driver with partly feedbacks, for ACTM or for NWP (data exchange with a limited time period coupling: off-line or on-line access coupling, with or without second iteration with corrected fields)
- Full chain of two-way feedbacks included on each time step (on-line coupling/integration)

There is a clear difference in needs for the on-line coupling of chemistry transport models and the off-line coupling. For the off-line coupled models it would advantageous to improve the quality of meteorological outputs, especially for fair weather conditions (including calm conditions and a focus on extreme situations). Parameters that can be used directly in off-line coupled models and that are important for processes like rainout should also be readily available. (It is important to remember that the couplings can be done in two directions: not only from NWP to atmospheric CTM, but also from atmospheric CTM to NWP, see above).

For the on-line coupling it is recommended to have a more modular setup of HIRLAM to make it easier to plug in the chemistry modules. Also the convection and condensation schemes need to be adjusted to take the aerosol–cloud interaction into account. Finally, the radiation scheme needs to be adjusted to include the aerosol effect.

In more details the advantages and disadvantages of the on-line and off-line modelling approaches are described in the Introduction.

The Enviro-HIRLAM (see its description and validation in a separate paper by Korsholm et al. in this volume) is a fully online NWP-ACT integrated system. The following steps towards Enviro-HIRLAM are being incorporated or have already been achieved: (1) nesting of models for high resolutions, (2) improved resolution of boundary and surface layer characteristics and structures, (3) 'urbanisation' of the model, (4) improvement of advection schemes, (5) implementation of chemical mechanisms, (6) implementation of aerosol dynamics, (7) realisation of feedback mechanisms, (8) assimilation of monitoring data (ongoing).

The Enviro-HIRLAM 10-year development history is outlined as the follows:

- 1999: Started as an unfunded initiative at DMI
- 2001: On-line passive multi-tracer pollutant transport and deposition mechanisms are included in HIRLAM-Tracer (Chenevez et al. 2004)
- 2003: Aerosol and chemistry model tested, at first, as 0D module in off-line CAC model (Gross and Baklanov 2004)
- 2004: Tests of different formulations for advection of tracers including cloud water (Lindberg 2004)
- 2005: Urbanisation of the meteorological model (within EC FP5 FUMAPEX project) (Baklanov et al. 2008)
- 2005: Study of aerosol feedbacks (COGCI, PhD study)
- 2006: Test of CISL advection scheme (Lauritzen 2005; Kaas 2008)
- 2007: First version for pollen studies (Mahura et al. 2009)
- 2008: New computationally efficient chemical solver NWP-Chem
- 2008: First version with indirect aerosol feedbacks (Korsholm 2009)
- 2008: Tests of new advection schemes in Enviro-HIRLAM (Univ. Copenhagen, PhD and MSc studies)
- 2008: Decision to build HIRLAM Chemical Branch (HCB) with the Enviro-HIRLAM as a baseline system
- 2008: Enviro-HIRLAM becomes the international project
- 2009: Opening of the HCB at http://hirlam.org
- 2009: Integrated version of Enviro-HIRLAM based on the HIRLAM reference version 7.2

Within the completely online Enviro-HIRLAM (Baklanov et al. 2008a; Korsholm et al. 2008b) the transport of chemical species is achieved in the same way like for other variables in HIRLAM (actually it was performed via the 'Tracer' subroutine inside HIRLAM like for other scalars) on the same time steps and with the same grid. There is no need for any interface in Enviro-HIRLAM, because ACT is inside the HIRLAM model, so all the HIRLAM parameters are available for ACT. The model is designed to be used for operational as well as research purposes and comprises aerosol and gas transport, dispersion and deposition, aerosol physics and chemistry, as well as gas-phase chemistry. A Climate version Enviro-HIRHAM is also planned and will be developed in the near future.

However, in many cases the classical way of the offline coupling/interfacing (or so called 'indirect way' when NWP and ACT are linked via an interface by input/output data exchange) is also very useful for different environmental forecast/assessment problems. The offline coupling of ACTMs and HIRLAM can be realised by dedicated interface modules (usually as a part of the ACTM) or using special 'couplers', e.g. European OASIS4 (see e.g. in Redler et al., in this Volume) or US ESMF (Dickenson et al. 2002). Most of the offline models of the HIRLAM community (e.g. CAC, EMEP, MATCH, SILAM) are using own interfaces. There was a small experience using OASIS4 by ECMWF for integration between NWP and ACTM within the GEMS project (Flemming et al. in this volume). However for *offline* or *online-access* modelling it is too computationally expensive, because 3D fields should be transferred on each time step (originally OASIS4 was developed for coupling of climate and the ocean models with 2D field exchange).

It is necessary to stress that the offline way of coupling is always more expensive in comparison with the online integration, because the reading, treatment and post-processing of NWP high-resolution meteorological fields for ACTM are very time consuming. It could be up to 80% of the total computational time (this is personal communications).

Further development and refinement of the online model coupling are expected in the future. It will lead to a new generation of integrated models for: climate change modelling, weather forecasting (e.g., in urban areas, severe weather events, etc.), air quality, long-term assessments of chemical composition and chemical weather forecasting. This way is much more advanced for considering aerosol and other precursors feedbacks on meteorological fields. The online coupling is also recommended when one wishes to improve the weather forecast (or climate) itself.

21.3 Improvements in Meteorology and Output

There are several shortcomings of the current HIRLAM and other NWP models for use in air pollution modelling and Chemical Weather Forecasting (CWF), including the following:

- Atmospheric environment modelling requires resolving more accurately the boundary and surface layer structure in NWP models (in comparison with weather forecast tasks).
- Despite the increased resolution of existing operational NWP models, urban and non-urban areas mostly contain similar sub-surface, surface, and boundary layer formulation.
- These do not account for specifically urban dynamics and energetics and their impact on the ABL characteristics (e.g. internal boundary layers, urban heat island, precipitation patterns).

- Numerical advection schemes in the meteorological part are not meeting all the requirements for ACTMs, so they should be improved and harmonized in NWP and ACT models.
- NWP models are not primarily developed for air pollution modelling and their results need to be designed as input to or be integrated into urban and meso-scale air quality models.

As stated above improvements in the quality of meteorological output require for off-line models especially in fair and calm weather conditions. Expected meteorological improvements in HIRLAM for ACT modelling include the following:

- PBL height (h), especially in very stable conditions (when h can be $<$ than the lowest model level) and over inhomogeneous surfaces (like urban or forest areas), where internal boundary layers play an important role and request using prognostic equations for PBL height (Zilitinkevich et al. 2007).
- Inclusion of urban characteristics (only for high resolution, important for stable conditions and weak winds).
- Improve surface drag over forest and city. Recent studies show that it may be twice as strong as currently in the models. Diagnostic estimation of U_{10m} and T_{2m} over urban areas should be improved.
- Inclusion of aerosol and cloud microphysics interaction.

Several steps in implementation of specific urban dynamics and energetics have already been achieved by the HIRLAM/HARMONIE community. Enviro-HIRLAM considers a surface improved description for urban areas: roughness, albedo, urban heat sources (Baklanov et al. 2008). Properties of urban aerosol used to modify the albedo characteristics and the effective radius of cloud droplets for the SW radiation (in the HIRLAM radiation scheme). In FUMAPEX two other more sophisticated urban schemes: BEP (Martilli et al. 2002) and SM2-U modules (Dupont and Mestayer 2006) were tested. They are more expensive computationally. Town energy balance (TEB) module (Masson 2000) is a part of SURFEX, available in the HARMONIE framework. Handling of the finest-scale details of momentum fluxes in town (forest) canopy could be developed.

Improvement of HIRLAM outputs for ACT modelling applications and correspondingly improvement of ACT models can include:

- Time averaged parameters (wind, temperature, cloud, precipitation, cloud top and bottom)
- Cloud top and bottom, Convective cloud top and bottom, Convective mass flux (if available)
- Stability parameters (Monin-Obukhov length, Deardorff convective velocity scale)
- PBL height as an output 2D field
- Availability of the same physiography database for surface, atmosphere and chemistry, depending on applications

21.4 Implementation of Chemistry, Aerosols and Feedback Mechanisms into Online Coupled ACTM-HIRLAM/HARMONIE

Some chemical species can also influence the weather, including:

- All greenhouse gases warm up the near-surface air.
- Aerosols: sea salt, dust, primary and secondary particles of anthropogenic and natural origin. Some aerosol particle components warm up and others cool down the air.
- Warm the air up (by absorbing solar radiation and thermal-IR radiation): black carbon, iron, and aluminium, polycyclic and nitrated aromatic compounds.
- Cool near-surface air down (by backscattering incident solar radiation to space): water, sulphate, nitrate, most of organic compounds.
- Different mechanisms of aerosols and other chemical species effect on meteorological parameters (direct, indirect effects, etc.).

Sensitivity studies are needed to understand the relative importance of different feedbacks. First experience of Enviro-HIRLAM indicates some sensitivity to effective droplet size modification in radiation and cloud characteristics (Korsholm et al. in this volume).

For the realisation of all aerosol forcing mechanisms in integrated systems it is necessary to improve not only ACTMs, but also NWP HIRLAM. The boundary layer structure and processes, including radiation transfer (Savijärvi 1990), cloud development and precipitation must be improved. Convection and condensation schemes need to be adjusted to take the aerosol–microphysical interactions into account, and the radiation scheme needs to be modified to include the aerosol effects. Incorporation of the aerosol direct effects (radiation forcing) is very problematic within the Savijärvi (1990) radiation scheme.

For simulation of the aerosol (especially anthropogenic) effects more detailed microphysics in HIRLAM is needed, in the former version it is very difficult to consider all the aerosol indirect effects. The current STRACO (Soft TRAnsition COndensation) scheme in HIRLAM (Sass 2002) needs modifications and gives a possibility of developing simpler indirect mechanisms. One of such feedback semi-empirical model was developed in Enviro-HIRLAM by Korsholm et al. (2008b). The new AROMA/HARMONIE cloud scheme (Pinty and Jabouille 1998; Caniaux et al. 1994) is more suitable for implementation of aerosol dynamics and indirect effects of aerosols (CCN) models, but will be more expensive computationally. So, the main focus of our collaboration in this field should be in the improvement of the cloud/microphysics and radiation schemes in HIRLAM.

The choice of the chemical and aerosol models for using with or implementation into HIRLAM/HARMONIE depends on specific tasks (e.g. for atmospheric pollution or for improvement of NWP). Current ACTMs of HIRLAM institutes (e.g. DACFOS, EMEP, MATCH, SILAM, etc.) are problem-oriented (air quality) and not very flexible for simple modifications. In CAC and Enviro-HIRLAM different chemical solvers, e.g. DMI module (Korsholm et al. 2008b), chemical module of WRF-Chem, RACM or RADM2, can be used. Other option could be the chemical mechanisms implemented into the latest version of the AROMA-C model (Tulet 2008). Our strategy in Enviro-HIRLAM is to build a universal cheap tool, which can build (automatically) a chemical solver for specific problem and requirements. It could lead to realization of a universal chemical module for HIRLAM. One of the possible options for that could be implementation of the KPP (Kinetic PreProcessor) tool (Damian et al. 2002) tested in the WRF-Chem system (Grell et al., this volume).

The aerosol module in Enviro-HIRLAM comprises two parts: (1) a thermodynamic equilibrium model (NWP-Chem-Liquid) (Korsholm et al. 2008a, b) and (2) the aerosol dynamics model CAC (Chemistry–Aerosol–Cloud) (Gross and Baklanov 2004) based on the modal approach. Other aerosol modules, including the sectional Model for Simulating Aerosol Interactions and Chemistry (MOSAIC) (Zaveri et al. 2007), the Modal Aerosol Dynamics Model for Europe (MADE) (Ackermann et al. 1998) with the secondary organic aerosol model (SORGAM) (Schell et al. 2001) (referred to as MADE/SORGAM) and the new Sectional Aerosol module for Large Scale Applications (SALSA) (Kokkola et al. 2008) are also expected to be used, but not tested and validated yet.

The path towards a coupled meso-scale HARMONIE-chemistry model is not clear yet. There is a possibility to use chemical components that have recently been included in AROME, but they have proved to be quite difficult to use. The Enviro-HIRLAM community is considering two different options for the development of a mesoscale coupled system: either the present HARMONIE chemistry code is used and extended with the chemistry and aerosol modules and feedback mechanisms that are already included in Enviro-HIRLAM, or Enviro-HIRLAM can be extended with an updated non-hydrostatic core NH-HIRLAM from Estonia (Rõõm et al. 2007).

21.5 Improvement of Advection Schemes

The default advection schemes used in HIRLAM have to be improved for ACT as well as for NWP and it should be a high priority task (see e.g. Baklanov 2008). For integrated ACTMs the requirements for advection schemes are even higher than for NWP models, they should be harmonised for all the scalars to keep the real mass conservation. So, in order to achieve the mass conservation, but at the same time to maintain large time steps for the solution of dynamical equations, the models (e.g. Enviro-HIRLAM) include several advection schemes (such as, semi-Lagrangian, Bott, SISL), which can be chosen in different combinations depending on specific problem. The easiest way for the online systems is using the same conservative scheme, e.g. the Bott scheme, for all the variables (e.g. for velocities, temperature, concentrations, etc.). But in comparison with the semi-lagrangian scheme of HIRLAM, the Bott scheme is four times more expensive. So, it is preferable to use the same conservative e.g., the Bott or cell-integrated semi-Lagrangian (CISL)

(Lauritzen 2005; Kaas 2008) scheme at least for chemicals, cloud water and humidity.

For offline ACTMs the choice of advection schemes is an independent and critically important problem. Additionally using NWP input data with non-conservative and not harmonized schemes (due to different schemes, grids, time steps in NWP and ACT models) offline ACTMs can get a dramatic problem with an explosive solution for the chemical part. For online coupling it is not a problem if one uses the same mass-conservative scheme for chemicals, cloud water and humidity.

It would be good to find an optimal advection scheme for integrated ACT-HIRLAM, however, estimations of ACT and NWP modellers can be different, because they can use different requirements/criteria for 'the best scheme' for NWP and ACTM. At least it is reasonable to analyze and compare different schemes used in the HIRLAM community (for NWP and ACT: including the semi-Lagrangian, CISL, Bott, Easter, Chlond, Walcek, Galperin and Kaas).

21.6 Importance of Data Assimilation and Problems with Data Assimilation for Chemistry

Data assimilation (DA) is important for daily chemistry forecasts, but it also is a challenge. At the moment there are much fewer stations with chemistry observations than meteorological observations, and the observations from these stations may not always be available in real time. Also, the gradients in chemical species often are very sharp, which the current observation network and data assimilation schemes are not capable of representing. And last but not least, it is difficult to determine the assimilation increments with so many possible species and so few detailed observations. This is why ECMWF has limited the chemistry in their model and data assimilation to five species (in addition to the cost in computer power).

The adjoint modelling technique is a good instrument to improve on the climatological emissions that are currently used for many species.

Data assimilation schemes for stand-alone atmospheric chemistry models are being developed by several groups in close collaboration with HIRLAM. Some of these schemes even have formulations close to the HIRLAM variational data assimilation. A coordination of these efforts for atmospheric chemistry data assimilation has a more long-term goal to extend the HIRLAM reference data assimilation to atmospheric chemistry.

21.7 Short-Term Perspectives and Long-Term Plans

In conclusion the above mentioned, we can say that ACT modellers expect from HIRLAM developers the following:

- Modifications of microphysics/clouds module (for online)
- Modifications of radiative and optical properties module (for online)
- Modifications of the cloud 3D data outputs (for offline)
- Improved description of PBL (SBL first of all), especially for urban, coast and forest areas (for all ACTMs)
- Continuity equation (advection) scheme improvement
- Open module structure of HIRLAM/HARMONIE

From other side the HIRLAM community should be interested in the improvement of NWP itself by implementation of ACTMs and aerosol/gases forcing/feedback mechanisms into HIRLAM (mostly by online integration), as well as by further development of the radiation and cloud microphysics modules to introduce there the feedback mechanisms. Besides, the above requested improvements of PBL structure in HIRLAM and its output will increase the number of possible HIRLAM applications for environmental protection and emergency preparedness modelling.

Proceedings from the conclusions of the section the following short-term perspectives and long-term plans can be outlined for the ACT and HIRLAM communities cooperation:

1. In the short term it may be possible to include the parameters that are requested and that may improve the offline coupling for ACTMs purposes. Here we are thinking about the average fields (accumulated) of wind components, temperature, specific humidity, cloud water and turbulent kinetic energy, and the parameters like 2D fields of cloud base and cloud top (to accurately determine where the scavenging is taking place). In addition some measures of stability (like the Monin-Obukhov Length, and PBL height) could also be included.
2. The vertical structure of PBL and SL is very important for ACTMs, so increasing the vertical resolution and improved parameterizations of BL are necessary. The meteorological improvements are already worked out for some aspects (stable PBL) and the urban characteristics should be included in the new surface scheme. In the meso-scale model the urban parametrization is already available (TEB: Town Energy Budget in SURFEX and BEP: Building Effect Parameterisation in Enviro-HIRLAM).
3. At the moment the cloud microphysics-aerosol interaction is included in HIRLAM in a very simple way in the convection schemes, where the cloud condensation nuclei have a lower concentration than over land. Enviro-HIRLAM includes the aerosol dynamics and their indirect effects on meteorology. The use of aerosol may also be prepared by making a 3D field of aerosol that has the characteristics of the currently prescribed values, then the extension to a real 3D distribution of aerosols that can interact with the microphysics is relatively straightforward. Sensitivity studies are needed to understand the relative importance of feedbacks. First experience of Enviro-HIRLAM indicates some sensitivity to effective droplet size modification in radiation and clouds.

4. Offline way of integration with reading ACTM output files by HIRLAM can also improve the NWP if aerosol feedback mechanisms are incorporated into HIRLAM. However, this way could be too expensive if high-resolution ACT forecast is available. For example, some tests can be done using the CAC, DERMA, EMEP, MATCH or SILAM models (considering feedbacks with Enviro-HIRLAM version). The OASIS4 coupler software (Redler et al., this volume) could be tested for this work.
5. Enviro-HIRLAM gives a good perspective for the future and it is a good candidate for inclusion in the reference system, however the online coupling to the HIRLAM model is a longer-term work. It may be quite straightforward if e.g. the Enviro-HIRLAM parts can be included in the reference as an option. It can only be included as an option because it may be too expensive for operational use by some partners. It would be good to start with a separate HIRLAM-CHEM branch.
6. Even the continuity equation (advection) scheme improvement is a more general problem in HIRLAM-A, it would be beneficial to coordinate this work of the HIRLAM 'Advection' team with Hirlam-countries ACTM developers (e.g., the CISL and other new schemes tests, inter-comparisons and implementation).
7. The Community has organised an initial working group (from HIRLAM and ACTM representatives from different HIRLAM-countries) for further HIRLAM-ACTM integration work, a sub-program (not completely inside the HIRLAM-A plan) for the Enviro-HIRLAM/HARMONIE (online and offline versions) development cooperation, and the HIRLAM Chemical Branch with the Enviro-HIRLAM model as the starting baseline online coupled system in HIRLAM.
8. A series of summer schools, workshops and training courses on "Integrated Modelling of Meteorological and Chemical Transport Processes / Impact of Chemical Weather on Numerical Weather Prediction and Climate Modelling" have been organised. They focused on the HIRLAM/HARMONIE-ACTM cooperation (the first one arranged 7–15 July 2008 in St. Petersburg, Russia, see: http://netfam.fmi.fi/YSSS08/). The aim of these events was to encourage the interaction between young scientists and researches interested in using and developing the HARMONIE (HIRLAM, ALADIN) modelling system, in order to elaborate, outline, discuss and make recommendations on the best strategy and practice for further developments and applications of the integrated modelling of both meteorological and chemical transport processes into the modelling system.

References

Ackermann IJ, Hass H, Memmesheimer M, Ebel A, Binkowski FS, Shankar U (1998) Modal aerosol dynamics model for Europe: development and first applications. Atmos Environ 32(17):2981–2999

Baklanov A (2008) Integrated meteorological and atmospheric chemical transport modeling: perspectives and strategy for HIRLAM/HARMONIE. HIRLAM Newsl 53:68–78

Baklanov A, Korsholm U (2007) On-line integrated meteorological and chemical transport modelling: advantages and prospective. In: Preprints ITM 2007: 29th NATO/SPS international technical meeting on air pollution. Modelling and its application. 24–28.09.2007, University of Aveiro, Portugal, pp 21–34

Baklanov A, Korsholm U, Mahura A, Petersen C, Gross A (2008a) ENVIRO-HIRLAM: on-line coupled modelling of urban meteorology and air pollution. Adv Sci Res 2:41–46

Baklanov A, Mestayer P, Clappier A, Zilitinkevich S, Joffre S, Mahura A, Nielsen NW (2008b) Towards improving the simulation of meteorological fields in urban areas through updated/ advanced surface fluxes description. Atmos Chem Phys 8:523–543

Caniaux G, Redelsperger J-L, Lafore J-P (1994) A numerical study of the stratiform region of a fast-moving squall line. Part I. General description, and water and heat budgets. J Atmos Sci 51:2046–2074

Chenevez J, Baklanov A, Sørensen JH (2004) Pollutant transport schemes integrated in a numerical weather prediction model: model description and verification results. Meteorol Appl 11(3):265–275

Damian V, Sandu A, Damian M, Potra F, Carmichael GR (2002) The kinetic preprocessor KPP – a software environment for solving chemical kinetics. Comput Chem Eng 26:1567–1579

Dickenson RE, Zebiak SE, Anderson JL, Blackmon ML, DeLuca C, Hogan TF, Iredell M, Ji M, Rood R, Suarez MJ, Taylor KE (2002) How can we advance our weather and climate models as a community? Bull Am Meteorol Soc 83:431–434

Dupont S, Mestayer PG (2006) Parameterization of the urban energy budget with the submesoscale soil model. J Appl Meteorol Climatol 45:1744–1765

Grell GA, Peckham SE, Schmitz R, McKeen SA, Frost G, Skamarock WC, Eder B (2005) Fully coupled "online" chemistry within the WRF model. Atmos Environ 39(37):6957–6975

Gross A (2000) Surface ozone and tropospheric chemistry with applications to regional air quality modelling. PhD thesis, University of Copenhagen and Danish Meteorological Institute. DMI Sci. Report No. 00-03. 296 p

Gross A, Baklanov A (2004) Modelling the influence of dimethyl sulphide on the aerosol production in the marine boundary layer. Int J Environ Pollut 22(1/2):51–71

Jacobson MZ (2006) Comment on "Fully coupled 'online' chemistry within the WRF model", by Grell et al. Atmos Environ 39:6957–6975

Kaas E (2008) A simple and efficient locally mass conserving semi-Lagrangian transport scheme. Tellus 60A:305–320

Kokkola H, Korhonen H, Lehtinen KEJ, Makkonen R, Asmi A, Jarvenoja S, Anttila T, Partanen A-I, Kulmala M, Jarvinen H, Laaksonen A, Kerminen V-M (2008) SALSA – a sectional aerosol module for large scale applications. Atmos Chem Phys 8:2469–2483

Korsholm US (2009) PhD thesis: Integrated modeling of aerosol indirect effects. Copenhagen University & DMI Scientific Report 09-01, www.dmi.dk/dmi/sr09-01.pdf

Korsholm U, Baklanov A, Gross A, Sørensen JH (2008a) Influence of offline coupling interval on meso-scale representations. Atmos Environ. doi:10.1016/j.atmosenv.2008.11.017

Korsholm US, Baklanov A, Gross A, Mahura A, Sass BH, Kaas E (2008b) Online coupled chemical weather forecasting based on HIRLAM – overview and prospective of Enviro-HIRLAM. HIRLAM Newsl 54:151–169

Korsholm US (2009) PhD thesis: Integrated modeling of aerosol indirect effects. Copenhagen University & DMI Scientific Report 09-01, www.dmi.dk/dmi/sr09-01.pdf

Lauritzen PH (2005) An inherently mass-conservative semi-implicit semi-lagrangian model. Ph.D. thesis, COGCI, University of Copenhagen, Niels Bohr Institute, Juliane Maries Vej 30, DK-2100 Copenhagen, Denmark

Lindberg K (2004) Using different formulations for the advection of tracers in DMI-HIRLAM. In 2 parts. DMI tech. reports No. 04-21 & 22

Martilli A, Clappier A, Rotach MW (2002) An urban surface exchange parameterisation for mesoscale models. Bound Layer Meteorol 104:261–304

Masson V (2000) A physically-based scheme for the urban energy budget in atmospheric models. Bound Layer Meteorol 98:357–397
Mahura A, Baklanov A, Korsholm U (2009) Parameterization of the Birch Pollen Diurnal Cycle. Aerobiologia 25:203–208
Pinty J-P, Jabouille P (1998) A mixed-phase cloud parameterization for use in mesoscale non-hydrostatic model: simulations of a squall line and of orographic precipitations. Proc. conf. of cloud physics, Everett, WA, USA, Amer. Meteor. soc., Aug. 1999, pp 217–220
Rõõm R, Männik A, Luhamaa A (2007) Nonhydrostatic semi-elastic hybrid-coordinate SISL extension of HIRLAM. Part I: numerical scheme. Tellus 59:650–660
Sass BH (2002) A research version of the STRACO cloud scheme. DMI Technical Report No. 02-10
Savijärvi H (1990) Fast radiation parameterization schemes for mesoscale and short-range forecast models. J Appl Meteorol 29:437–447
Schell B, Ackermann IJ, Hass H, Binkowski FS, Ebel A (2001) Modeling the formation of secondary organic aerosol within a comprehensive air quality model system. J Geophys Res 106:28275–28293
Tijm S, Baklanov A, Rontu L (2007) HIRLAM: air chemistry transport modelling and numerical weather prediction. Conclusions from HIRLAM/HARMONIE-ACT models integration session, COST728-NetFAM workshop, 28.5.2007. http://netfam.fmi.fi/Integ07/HARMONIE_followup.doc
Tulet P (2008) AROMA-C. User manual. MeteoFrance, Toulouse
Zaveri RA, Easter RC, Fast JD, Peters LK (2007) Model for simulating aerosol interactions and chemistry (MOSAIC). J Geophys Res. doi:10.1029/2007JD008782
Zilitinkevich S, Esau I, Baklanov A (2007) Further comments on the equilibrium height of neutral and stable planetary boundary layers. Q J R Meteorol Soc 133:265–271

Chapter 22
Summary and Recommendations on Integrated Modelling

Alexander Baklanov, Georg Grell, Barbara Fay, Sandro Finardi, Valentin Foltescu, Jacek Kaminski, Mikhail Sofiev, Ranjeet S. Sokhi, and Yang Zhang

22.1 Introduction

This Chapter summaries the main discussion points arising from the topics of the workshop, namely:

- On-line and off-line coupling of meteorological and air quality models
- Implementation of feedback mechanisms, direct and indirect effects of aerosols
- Advanced interfaces between NWP and ACTM models
- Model validation studies, including air quality-related episode cases

It also draws together some conclusions that have strategic implications in this research area.

In order to review progress in off-line and on-line coupled models it is important to have exchange of experience from a wide range of groups involved in the development and application of these systems. The sections summarises inputs from a number of groups from USA, Canada, Japan, Australia, and Europe. It is hoped that these discussions will lay the foundation for a framework for an integrated mesoscale modelling strategy for Europe.

The round table discussion inevitably examined the main requirements of such a strategy. The points discussed included:

- The large number of models and applications within Europe
- The need for modelling communities to have closer interaction (such as among air quality, NWP, and climate communities and between model developers and users)
- The important science questions to be addressed
- The implications for users and for operational applications
- Flexibility of the strategy to allow for the diversity in European modelling community

A. Baklanov (✉)
Danish Meteorological Institute (DMI), Lyngbyvej 100, DK-2100 Copenhagen, Denmark
e-mail: alb@dmi.dk

A. Baklanov et al. (eds.), *Integrated Systems of Meso-Meteorological and Chemical Transport Models*, DOI 10.1007/978-3-642-13980-2_22,
© Springer-Verlag Berlin Heidelberg 2011

- Inclusion of off-line and on-line systems
- The importance exchanging experience and knowledge between international groups
- Any limitations of such a strategy

22.2 On-Line and Off-Line Coupling of Meteorological and Air Quality Models

The aim of this section is to pull together the main points on "coupling" of Numerical Weather Prediction and/or Meso-meteorology (NWP/MM) and Atmospheric Chemical Transport Model (ACTM). Where possible, recommendations are given on the best practice and strategy for further developments and applications of integrated modelling systems.

The following types of coupling can be considered:

- Off-line with no or very limited coupling
- On-line access (with availability of meteorological data at each time step)
- On-line integrated (with feedbacks possible to consider from ACTM to NWP/MM)

Feedback mechanisms are considered to be important for a coupling strategy. This is especially relevant for aerosols. Aerosol forcing mechanisms influence radiative and optical properties as well as cloud processes, leading for instance to changes of precipitation and circulation.

The strategic recommendations relating to on-line coupling and feedbacks include the following:

- Need to identify key examples of significant feedbacks that were identified so far in sensitivity studies performed with on-line systems
- Communicate and encourage exchange of research and development information within the NWP/MM modelling community on key feedback processes and their magnitudes

Model interfaces include many aspects of the interoperability of NWP/MM and ACTM models. Modular coding is advocated in order to ease implementation of different algorithms/routines serving the same purpose. Ensuring common standards for data exchange are also important.

Recommendations are given below and these complement those from other sections:

- In order to foster collaboration it is important to encourage the sharing of modules
- Further study is required to develop and improve coupling interfaces dealing specifically with chemical-data assimilation
- The computational efficiency when using different couplers needs to be investigated

- Guidance is required on when to employ time-averaged values and instantaneous values of meteorological variables driving the ACTM routines
- Continue to highlight opportunities to develop further coupling interfaces between NWP/MM and ACTM

Other points which need more attention in respect to "coupling" NWP/MM and ACTM models involve:

- Inclusion of the sea breeze process which is crucial for chemical weather modelling in coastal regions. The meteorological driver needs to describe properly the processes relevant to sea breeze circulations.
- Improved parametrization of urban effects on the atmospheric boundary layer (BL) is needed. NWP/MM models despite their increased resolution, still have shortcomings. For instance, the description of sub-surface, surface and urban BL for urban areas is similar to that of rural areas. Thus, the urban dynamics and energetics are not properly described. NWP/MM models are not primarily developed for air pollution modelling, and their outputs have to be made suitable to provide input for urban-scale ACTMs.

22.3 Implementation of Feedback Mechanisms, Direct and Indirect Effects of Aerosols

One of the important tasks is to develop a modelling tool of coupled 'Atmospheric chemistry/Aerosol' and 'Atmospheric Dynamics/Climate' models for integrated studies, which is able to consider feedback mechanisms, such as, aerosol forcing (direct and indirect) on the meteorological processes and climate change.

Chemical species influencing weather and atmospheric processes include greenhouse gases which warm near-surface air and aerosols such as natural aerosols (e.g., sea salt, dust, volcanic aerosols) and primary and secondary particles of anthropogenic origin. Some aerosol particle components warm the air by absorbing solar radiation (e.g., black carbon (BC), iron, aluminium, polycyclic, and nitrated aromatic compounds) and thermal-IR (e.g., dust) , whereas others (water, sulphate, nitrate, most of organic compounds) cool the air by backscattering incident short-wave radiation to space.

It is necessary to highlight the effects of aerosols and other chemical species on meteorological parameters, which have many different pathways (direct, semi-direct, and indirect effects, etc.) and they have to be prioritised and considered in on-line coupled modelling systems. Sensitivity studies are needed to understand the relative importance of different feedback mechanisms for different species and conditions relevant to air quality and climate interactions. A concerted action to mobilise and coordinate research in this area is needed.

A particular area of development for on-line coupled models is in the area of parametrizations that allow for interactions of physics with chemistry. With the realization of the increasing importance of science questions related to global and regional climate change, two-way interactions between meteorology and chemistry are becoming a necessity in complex 3D models. Complex 3D models are increasingly being used not only for meteorological predictions, but to better understand and simulate the wide range of processes and atmospheric feedbacks that influence climate. In contrast to global climate models, the flexible grid structure of high resolution nonhydrostatic models enables the simulation of climate processes at spatial and temporal scales compatible with measurements, providing a framework in which to test new parametrizations of climate processes that are either treated in a simple way in current global climate models or neglected entirely.

The coordination of parametrizations that allow for interactions between aerosols, radiation, chemistry, and clouds, as well as the coordination of new treatments for meteorological process modules (e.g., boundary layer, clouds) that are needed to improve predictions of atmospheric chemistry is an important consideration. Model development in atmospheric chemistry and weather prediction has so far developed separately from each other, leading, for example, to meteorological parametrizations that have no treatment of any chemical species. A well known example of linkage between chemistry and meteorology includes the treatment of cloud–aerosol interactions. Most current cloud physics schemes neglect the linkages of CCN to predicted aerosol distributions. Coupling the chemistry part with cloud physics involves modifying existing chemistry and cloud subroutines that normally do not interact with each other. This is a very complicated process that requires considerable effort.

One example not related to the aerosol direct or indirect effects is the current treatments of boundary layer mixing which greatly affects near-surface concentrations. Quite often chemical models (if running on-line) rely on eddy coefficients from the meteorological model for the vertical mixing of trace-gas and particulate scalars. This only works since the mixing is determined by eddy coefficients that are calculated in the meteorological part and can then be used in the chemical part to mix the tracers. It cannot work if the meteorological parametrization uses a different method to mix tracers. It would be much more desirable if designers of the meteorological parametrizations would consider the mixing of a scalar in their scheme from the beginning. This would lead to more general and more accurate vertical mixing algorithms that handle both meteorological and chemical scalars. In addition it would provide meteorological modelers with an additional independent source for possible evaluation. Similar arguments can be made for the treatment of parameterized convection, where only very few convective parametrizations exist in meteorological models that allow for transport and modification of tracers.

Implementation of the feedbacks into integrated ACTM-NWP models could be realized in different ways with varying complexity. The following variants serve as examples:

- Simplest off-line coupling: The chemical composition fields from atmospheric CTMs may be read by MetM/NWP at a limited time period and used as driver for aerosol forcing on meteorological processes.

- On-line access coupling: Driver and partial aerosol feedbacks, for ACTMs or for NWP (data exchange on each time step) with or without the following iterations with corrected fields.
- Fully on-line coupling/integration: ACTM and feedbacks included inside MetM for each time step.

The above examples represent the different levels of on-line integration that have been achieved within Europe. Many of the systems are currently either fully off-line or have some on-line capabilities. A few systems now exist that are fully on-line coupled (e.g., ENVIRO-HIRLAM, WRF-Chem and UKCA systems). Historically Europe has not adopted a community approach to modelling and this has led to a large number of model development programmes, usually working independently. However, a strategic framework will help to provide a common goal and direction to European research in this field while having multiple models.

There are a number of key elements that need to be part of the overall strategy framework.

These include:

1. Scientific questions to be addressed by on-line systems

As stated earlier the major reason for developing on-line systems is to take account of feedback mechanisms for accurate modelling of NWP/MM-ACTM and quantifying direct and indirect effects of aerosols. Several questions can be identified in this regard:

- What are the effects of climate/meteorology on the abundance and properties (chemical, microphysical, and radiative) of aerosols on urban/regional scales?
- What are the effects of aerosols on urban/regional climate/meteorology and their relative importance (e.g., anthropogenic vs. natural)?
- How important the two-way/chain feedbacks among meteorology, climate, and air quality are in the estimated effects?
- What is the relative importance of aerosol direct and indirect effects in the estimates?
- What are the key uncertainties associated with model predictions of those effects?
- How can simulated feedbacks be verified with available datasets?

2. Processes/feedbacks to be considered

A detailed treatment of the main processes is required in the models in order to answer the above questions. These processes include:

- *Direct effect* – decrease solar radiation and visibility:
  - Processes needed: radiation (such as scattering, absorption, and refraction)
  - Key variables: refractive indices, extinction coefficient, single scattering albedo (SSA), asymmetry factor, aerosol optical depth (AOD), and visual range
  - Key species: cooling: water, sulfate, nitrate, and most organic carbon (OC) warming: black carbon (BC), OC, iron, aluminium, and polycyclic/nitrated aromatic compounds

- *Semi-direct effect* – affect planetary boundary layer (PBL) meteorology and photochemistry:
  - Processes needed: PBL/land-surface (LS), photolysis, and other met-dependent processes
  - Key variables: temperature, pressure, relative humidity, water vapor mixing ratio, wind speed, wind direction, cloud fraction, stability, PBL height, photolysis rates, and the emission rates of met-dependent primary species (e.g., dust, sea-salt, biogenic aerosol, marine phytoplankton-produced aerosol)
- *First indirect effect* – affect cloud drop size, number, reflectivity, and optical depth via cloud condensation nuclei (CCN):
  - Processes needed: aerosol activation/resuspension, cloud microphysics, and hydrometeor dynamics
  - Key variables: interstitial/activated fraction, CCN size/composition, cloud drop size/number/liquid water content (LWC), cloud optical depth (COD), and updraft velocity
- *Second indirect effect* – affect cloud LWC, lifetime, and precipitation:
  - Processes needed: in-/below-cloud scavenging and droplet sedimentation
  - Key variables: scavenging efficiency, precipitation rate, and sedimentation rate
- *All aerosol effects*:
  - Processes needed: aerosol thermodynamics/dynamics, aqueous-phase chemistry, gaseous precursor emissions, primary aerosol emissions, and water uptake
  - Key variables: aerosol mass, number, size, composition, hygroscopicity, and mixing state.

3. Implementation Strategies and Milestones

A crucial part of implementing a strategy is to have workable objectives that span short and long term time frames. Some of the elements that will need to be considered include:

- Support strategically both off-line and on-line coupled model frameworks
- Identification of the capabilities of on-line models for air quality applications
- Strategy should be implemented in a phased manner to allow groups to opt in depending on their requirements, capabilities, and needs
- Development of benchmarks or guidelines should be considered reflecting the main uses and applications for Europe
- In light of these developments computational resources available in Europe for intense computational studies should be re-assessed and future requirements identified
- Where possible common model evaluation including design of process-based field and laboratory studies (e.g., closure experiments) should be suggested
- An important aspect would be to initiate and encourage the training of graduate and post-doctoral researchers via formal courses/summer schools

4. Coordination Plan and Logistics

COST-728 Action has been instrumental in laying the first foundations of a European-wide strategy. However, for the strategy to have wide acceptance, close cooperation and interaction will be required with the main groups in Europe. The following international coordination activities will need to be established:

- Progress review meetings annually
- Establish working subgroup focusing on specific areas
- Annual workshop alternating in Europe and North America
- Working group meeting in conjunction with annual workshop
- Special journal issues dedicated to workshop papers
- Plan for securing resources and funding to implement the strategy that is ambitious, forward-looking, and sustainable

22.4 Advanced Interfaces Between NWP and ACTM Models

There are a number of points that need to be considered when developing new integrated model systems. The basic recognition is for the importance of dealing with the requirements of the chemical model at the same time as with the meteorological ones. In the past, this was normally neglected and the system designed on meteorological grounds while only adding in the chemistry later. With modern supercomputers, on-line coupled systems are decidedly preferable, guaranteeing high temporal resolution of all coupled processes and feedbacks. There are several advantages of including many models as modules into a common system framework for maximum information, flexibility, and applicability. At the same time, computer resources should allow modelers to be free of needs for parallelization or other cumbersome technical requirements.

Internationally there are several on-line systems. In the U.S., these systems include GATOR-GCMOM, WRF/Chem, and several variants of WRF/Chem such as WRF/Chem-MADRID and global-through-urban WRF/Chem (GU-WRF/Chem) for applications from global to urban scales. In Australia, an on-line system that is employed at CSIRO for Australian applications works well on the smaller scales (1–1,000 km), e.g., for the import/export of bush or forest fire smoke, while the system is used at the UK Hadley Centre for the larger, global scale. In Japan an adapted version of WRF/Chem model is being used. Off-line models used in Japan are mainly US models for chemistry while the meteorological models are provided by the Japanese meteorological agency. Within Europe examples include of the on-line coupling of the COSMO-EU (formerly Lokalmodell) to chemistry model. The German Weather Service DWD and the MPI for Meteorology in Hamburg are developing the ICON model as a combined NWP and climate model on a scale-adaptive icosahedral grid including chemistry to become the operational NWP model at the DWD in 2012. There was a prime interest in precipitation forecasting at ECMWF, and the current introduction of chemistry into the models (in GEMS) in order to build an integrated forecast system. The HIRLAM model now has on-line chemistry features at DMI (see Chapter 21).

Given the historical diversity of the model development trail in different countries of Europe, a single system approach will not be a viable option but a common framework which can eventually develop into an open system platform for partners to add modules with harmonized interfaces and parameters could be considered (e.g., an open system that is similar to WRF/Chem in the U.S.), but will still require improved communication and cooperation between the European partners.

Taking into account the large number of smaller institutes and administrations across Europe engaged in air quality modeling, the need for practical, easy-to-handle solutions on small computers with few staff and restricted modelling experience has been emphasized at this and other meetings. These systems are of necessity off-line often using measurement NWP forecasts of ECMWF or national met offices. Therefore, these systems often strongly depend on interfaces connecting the chemistry modules to measurement post-processing and/or meteorological modules.

The need for unification of interface modules at least concerning high modularity and I/O format specifications, but possibly also parametrizations and for according guidelines was also apparent. Lessons should be learned from ECMWF practice to develop European guidelines.

22.5 Model Validation Studies, Including Air Quality Related Episode Cases

A problem of coupling the NWP and ACTM models does not necessarily follow the line between on- and off-line methods of coupling. Numerous compromises accepted in both models, different histories and tasks have created several niches for various combinations of these systems and for various interfaces between them. For example, primarily off-line interface between the meteorological and dispersion models is used for real-time emergency applications. Huge uncertainties in information about the source term in any real emergency situation and strong time limitations for analysis make cheaper and faster off-line models the only practical choice. On the other hand, an extreme example is climate-related studies where excluding the interaction between atmospheric composition and dynamics is hardly acceptable. Most other applications are somewhere in-between the above extremes and a particular choice of modelling tools and their interactions vary from case to case.

It is worth mentioning that the reasons for moving towards more complicated but also more comprehensive on-line coupled systems can also be different, as can be the actual means of the on-line coupling. For instance, the access to NWP fields at each time step is clearly an advantage providing that the dynamic core is unique for both meteorological and chemical parts. Otherwise, differences in features of the advection and diffusion schemes of the coupled models will create tough problems instead of solving them. The meteorological fields computed within the NWP sub-model with own advection implementation will be inconsistent from the point of view of chemical transport sub-model. Most modern ACTMs have special meteorological pre-processors that deal with this inconsistency, which is then treated as a feature of the NWP fields. To overcome this very serious obstacle, the NWP and ACTM models have to be actually unified, so that a single dynamic core handles both chemical and meteorological transport. A pre-requisite for this is a decent dynamics, which satisfies both sides – a luxury rarely available in existing modelling systems, especially if they were created separately.

The other strong motivation to move towards on-line coupling is a possibility to simulate feedbacks from chemical part to meteorological one. Being indeed simpler in such a model configuration, the feedbacks themselves can be comparatively easily arranged also with off-line systems with very limited computational overhead. Admittedly, the off-line feedbacks can hardly reach the frequency of every model time step; but in most cases this is not needed either. The feedback impact even in strongly polluted areas is still a relatively small perturbation for the main processes. Nearly the only exception is the dust storm events where the aerosol concentration is so high that it cannot be considered as a tracer any more. A simplification, however, is that it is enough to add just a couple of aerosol modes into the NWP variables while still keeping the bulky chemistry as an external model.

In conclusion, it is worth recognizing that the advantages and disadvantages of the on-line and off-line coupling of the meteorological and chemical transport models can depend on the specific task and objective of the applications. Diversity and specifics of the tasks and objectives, in their turn, will require the existence of both approaches for the foreseeable future.

Finally, it is important to highlight that this initiative to develop a European strategy for on-line coupled mesoscale models for air quality applications should fully benefit from all relevant groups and projects including COST 728 and other initiatives such as COST ES0602 (Towards a European Network on Chemical Weather Forecasting and Information Systems). It is also vital that the inputs of users and policy makers are given prominence and accommodated in forging the future scientific research direction of this field.

Index

A
ACCESS and HADGEM, 143
Acidifying compounds, 148
Adjoint modeling, 207
Advection, 115, 117, 119, 121
Advection schemes, 218, 220, 222–225
Aerosol, 16–20, 27–32, 55, 57, 58, 82, 148, 207–209
  -climate feedbacks, 15, 33
  -cloud interactions, 20, 30–32
  dynamics, 78, 79, 172
  feedback mechanisms, 225
  feedbacks, 9, 47
Air pollution forecasts, 175
Air quality (AQ), 55–57, 81–87
  boundary conditions, 100, 105, 106
  forecasting, 142, 147–152
  initial conditions, 100, 105
  modeling, 97–107
AirQUIS, 147–151
ALADIN, 195–196
ALADIN-CAMx interface, 195
Alert/information threshold, 197–199
Anthropogenic emissions, 59
Atmospheric boundary layer (ABL), 56
Atmospheric chemistry mechanisms, 81
Atmospheric forcing, 202
Atmospheric transport, 148, 157, 165
Atmospheric Transport Model Evaluation Study (ATMES), 55, 65, 66
Australia, 139–143
Austrian Air Quality measurements, 196
Austrian Air Quality model, 195

B
Biogenic emissions, 59, 76
Biomass burning, 183, 185, 188
BOKU, 195
BOLam +CHEMistry, 89
BOLogna Limited Area Model (BOLAM), 89, 90
Bott advection scheme, 63
Boundary conditions, 201, 202, 205
Boundary layer parameterization, 98
Boundary layer parameters, 149, 151, 152
Building wake, 140
Bushfires, 142

C
CAMx, 195, 199
Carbon monoxide (CO), 59, 182–183, 185–190
Cell Integrated Semi Lagrangian (CISL), 63
CHASER, 181–192
Chemical branch, 218, 225
Chemical equilibrium, 120
Chemical transport model (CTM), 16, 20–22, 61, 63, 72, 113–117, 119, 121, 122, 147–149
Chemical weather forecasting (CWF), 1, 4, 9, 42, 55, 172, 175–176
Chemistry-meteorology two-way interaction, 9, 42
Chemistry schemes, 170
Chemistry tendency, 112, 115–119, 121, 122
Chernobyl, 66, 68, 73
Climate modelling, 125, 126, 129, 132
Cloud microphysics, 140
Clouds, 15, 17–19, 27
Commonwealth Scientific and Industrial Research Organisation (CSIRO), 139
Communication, 133–135
Community network, 129, 130
Computational performance, 114–115
Condensed chemistry model, 140
Constant boundary conditions, 196–197
Continental outflow, 181
Convection, 117–119, 121, 122

A. Baklanov et al. (eds.), *Integrated Systems of Meso-Meteorological and Chemical Transport Models*, DOI 10.1007/978-3-642-13980-2,

© Springer-Verlag Berlin Heidelberg 2011

Convective transport, 56, 59, 181, 187–190
COSMO-ART, 75–79
COST Action 728, 1–2
Coupled ACTM-HIRLAM, 221–222
Coupler, 125–128, 132–135
Coupling, 109–122
Cumulus parameterization, 152

D
Data archiving, 128
Data assimilation, 223
Data exchange, 133
Data networking, 128
Data processing, 128
Diffusion, 116–119, 121, 122
Dislocation, 112, 115–116, 119
Dispersion, 62, 64–67, 69, 71, 73
  modeling, 160
  parameters, 149, 150
DMI ACT models, 167–176
Downscaling, 143
Dry deposition, 58, 63, 68
Dust emissions, 76
Dynamic boundary conditions, 197

E
Earth system models, 129–131
East Asia, 181
ECHAM5, 131
ECMWF, 109–122, 207–212
ECMWF total ozone column, 196
Effects of interface nesting, 100–102
EMEP emissions, 196
EMEP model, 148, 149, 151–152
Emission, 147–148
  injection, 116, 119, 122
  scenarios, 82, 87
Ensemble modeling, 208
Entrainment/detrainment rate, 152
Enviro-HIRLAM, 4–6, 61–73, 215–218, 221–222, 224–225
Eulerian model, 209
European strategy for integrated modeling, 235, 237
European Tracer Experiment (ETEX), 64–66, 69, 73
Eutrophying compounds, 148
Evaluation, 198
Explicit rain processes, 140

F
Feedback, 61, 62, 71–73
Feedback mechanisms, 229–235
Flux transport scheme, 155
Forecast evaluation, 149–151
Forward modeling, 207
Framework, 125, 127, 130, 135, 136

G
Gas phase, 55, 57–58
GEMS, 135
Global statistics, 68
GME model, 76
Ground-based observation, 185–186, 190

H
HARMONIE, 215–225
Highly resolved emission data, 78
Highly resolved emission inventory, 196
High Resolution Limited Area Model (HIRLAM), 61–73, 207–212, 215–225
Hit-rate of exceedances, 198–199

I
Initialization of wind data, 155–157
Integrated modeling system, 5, 10, 42
Intercontinental transport, 181, 184

J
Japanese Air Quality Model, 190, 191
JCAP emission inventory, 190–191
J values, 57, 58

L
Lagrangian model, 208
Lagrangian particle model, 139, 147
Lapland, 207–212

M
Mass-conservation, 155–157
Mesoscale climate chemistry model (MCCM), 81–87
Mesoscale meteorology, 139
Mesoscale transport and stream (METRAS), 201–205
Mesoscale variability, 62
MESSy, 131
Metadata, 128–129, 136
Meteorological coupling interval, 68–71, 73
Meteorological modeling and forecasting, 229–232, 235
Meteorology-chemistry integrated models, 232, 235
Mexico City, 85
MM5, 147–151

MM5/chem, 82
Model infrastructure, 130
Model intercomparison, 173
Model interface, 151
Modelling framework, 130–131
Model mechanism comparison, 19
Model validation, 164–165
M-SYS, 201, 202

N
Nesting, 83, 86, 201–206
Nitrogen oxides (NO_2), 59, 147, 149–151
Non-hydrostatic, 81, 147, 151
Non-hydrostatic *vs.* hydrostatic input data, 156, 157
NOx-inter-conversion, 119–121
Nuclear emergency system, 147
Numerical weather prediction (NWP), 1–4, 8, 9, 42
NWP-CHEM, 63

O
OASIS, 125–137
Off-line, 61–73
  AQ models, 97, 98
  coupling, 148–149, 151
  coupling meteorological, 97, 106
  model, 155, 165, 195
One-way nesting, 191
On-line, 61–73
  coupling, 17, 19, 75
  integration, 110
  model, 182
  and off-line coupling, 1–10, 41, 42, 217–219
  vs. off-line coupled models, 237
Online coupled meteorology-chemistry model, 81
Operational, 147, 148
Operational applications, 229
Operator splitting, 116
Ozone, 57
  forecasts, 191–192, 195–199
  sensitivity, 91–92

P
Passive tracers, 173
Peak concentration, 209
Personal exposure, 142
Photochemical air pollution, 140
Photodissociation rates, 57
Photolysis, 120
Photooxidants, 87
Photo-oxidants, 148
Physical and chemical weather, 9
Plume rise, 140
PM2.5, 147
PM10, 147
Point source, 207, 211
Pollen, 76
Pollutant plume, 209

R
Radiative impact, 75–77, 79
RADM, RACM, RACM-MIM, 82
REAS emission inventory, 190–191
Regional air pollution, 140
Regional climate chemistry simulations, 83
Regridding, 133–135
Rome, 101, 102

S
Saharan dust transport, 92
Scavenging, 58
Sea breeze, 140, 141
Seasalt emissions, 76
Semi-Lagrangian, 63
Semi-Lagrangian advection, 56, 57
Short-range forecast, 61, 69
SILAM, 207–212
SNAP model, 147, 148
Source receptor analysis, 86
Stratosphere, 59
Sulphate, 207–212
Surface classes, 151–152

T
The Air Pollution Model (TAPM), 139–143
Torino, 104–106
Transport, 109–122
Troposphere, 57, 59
Tropospheric chemistry, 55, 82

U
UK Unified Model, 151
Update intervals of meteorological forcing, 201
Urban air pollution, 140, 141
Urban air quality, 147, 149–151
Urban parameterization, 169, 172–174

V
Värriö monitoring station (SMEAR I), 207
Vertical diffusion, 56, 58

W
Weather forecasting, 55
Weather Research Forecast model with Chemistry (WRF-Chem), 41, 44, 45, 47–51, 181–192
Well-mixing assumption, 209
Wet deposition, 63, 67, 68
Wind-blown dust, 142
Winter time inversions, 147

Z
ZAMG, 195, 196